AF879700

Corporate Governance in Commonwealth Countries

Edited by

Elewechi Okike

Centrag

International Centre for Research in Accountability and Governance

Copyright © 2019 Elewechi Okike

British Library Cataloguing in Publications Data

All rights reserved. No part of this work may be reproduced or stored in an information retrieval system (other than for purposes of review) without the express permission of the publisher in writing.

A catalogue record for this book is available from the British Library.

ISBN: 978-1-9160282-1-0

A Publication of the International Centre for Research in Accountability and Governance,

P.O. Box 112, Washington NE37 1YB, United Kingdom

Tel: 07799653641

Email: info@centrag.org

http://www.centrag.org

All efforts have been made to trace and acknowledge ownership of copyright. The Publisher would be glad to hear from any copyright holders whom it has not been possible to contact.

Although the information provided in this publication is accurate at the time of publication (January 2019), the subject is constantly evolving and updates will be provided in subsequent editions.

The publisher accepts no responsibility for any acts or omissions contained, and cannot guarantee the accuracy and completeness of the contents of contributions by authors.

Views expressed in this publication do not necessarily reflect those of the publisher.

This book is dedicated to the
glory of His name, Who, against all
odds made it possible.

Acknowledgements

I am grateful to God for the grace and the tenacity to complete this book, which is a publication of the International Centre for Research in Accountability and Governance (CENTRAG). I wish to appreciate all the authors who have contributed chapters to the book, especially for their patience, as the publication took much longer than had been anticipated. Special thanks also to the academic reviewers, who provided valuable feedback on individual chapters. Dr Chikezie Okike and other colleagues at CENTRAG did a great job in providing editorial support for the book. I am grateful to them and to Susan M. for putting the book together.

CONTENTS

Contributors

Ismail Adelopo

Bristol Business School, University of the West of England, UK

Prof Ismail Adelopo is an associate professor in Accounting and Finance at the University of the West of England, Bristol, UK. His research focuses on the intersection between corporate governance, accounting and auditing. He sees and uses accounting and finance as an instrument of change and explores the role of governance in this. The other aspect of his research explores corruption and anti-corruption initiatives. In particular he studies the factors that inhibits the success of anti-corruption measures in countries with systemic corruption.

Beenish Ameer

Financial and Corporate Integrity Centre, Coventry University, UK

Beenish Ameer is a PhD student at the Financial and Corporate Integrity Centre, at Coventry University. She graduated as the top MBA student at Quaid-I-Azam School of Management Sciences, Quaid-I-Azam University, Islamabad, Pakistan, and was awarded the Vice Chancellor's talent scholarship. She has investigated a number of emerging issues in Pakistan, including examining the job satisfaction level of academic staff in Pakistani universities, the borrower perspective of the effectiveness of microfinance loan in Pakistan, the Fama and French three factor model in Pakistan Equity Market and corporate governance issues and challenges in Pakistan, among others. She is widely published and sits on the editorial board of *Global Journal of Management and Business Finance.*

Fadeke Ayoola

Natural Environment Technology in Africa, Montgomery House, Belgium

Fadeke Ayoola is the CEO of Natural Environment Technology (NET) Africa, Chartered Certified Accountant (FCCA), Author, Lead Environmental Auditor ISO 14001:2015 and Visiting Lecturer. Her current research interests include a critical review of the regulatory framework of Environmental Impact Assessments (EIA) within the countries of the African Great Lakes. Her fieldwork and projects include the promotion of sustainable management of wetlands of trans-boundary significance, to protect sustainable management of critical water source catchments through environmental management and audit techniques.

Ibrahim Bedi

Department of Accounting, University of Ghana Business School, Accra, Ghana

Dr. Ibrahim Bedi is a chartered accountant and a Senior Lecturer at the Department of Accounting, University of Ghana Business School. He is a fellow of the Association of Certified Chartered Accountants (FCCA) (UK); member of the Institute of Chartered Accountants, Ghana; member of American Accounting Association; treasurer of African Accounting and Finance Association; and member of International Association of Accounting Educators and Researchers. He holds a Doctor of Philosophy, Master of Business Administration and Bachelor of Science in Administration in accounting degrees. In 2010, he was a visiting scholar to the Open University, UK. He is a partner of Taylor, Folson and Associates (Chartered Accountants). His research interests include corporate governance, reporting quality, IFRS compliance and disclosure, audit quality, audit cost and cost of capital, changes in accounting education and SDG 17 as well as tax compliance and earnings management.

Samuel E. Fulgence

De Montfort University, Faculty of Business and Law, Accounting and Finance Department, UK

Samuel Fulgence is a Ph.D candidate in the Faculty of Business and Law at De Montfort University. He received his MSc in Accounting and Finance with distinction from London Metropolitan University, UK, for which he won an award for academic excellence, as the Best Master's student. He is a registered Tax Consultant with Tanzania Revenue Authority (TRA) and an Associate Certified Public Accountant in Public Practice (ACPA-PP). Before embarking on the Ph.D programme, Samuel was a lecturer, Consultant and Head of the Department of Accounting and Finance at Tanzania Institute of Accountancy. He also worked as Senior Audit Manager at SIBE Financial Services and Co-Consultant with Kikoti Financial Consult. He is an Editorial Board Member for *Global Scholars Journal of Accounting and Taxation* (GSJAT) and a resource person at the National Board of Accountants and Auditors (NBAA) in Tanzania. He is an approved IPSASs trainer and an expert in Corporate Governance and Public Sector Accounting and Finance. His research interests include corporate governance and capital structure in which he has authored several books, book chapters and peer reviewed articles.

Andrew Munthopa Lipunga

Department of Accountancy, University of Malawi – The Polytechnic, Blantyre 3, Malawi

Andrew Munthopa Lipunga is the Deputy Head of Accountancy and Senior Lecturer in the Department of Accountancy at a constituent college of the University of Malawi. He teaches financial reporting, financial management and corporate governance and ethics. His research interests are in corporate governance practices in developing countries and corporate reporting and in particular focusing on non-financial disclosures.

Md Humayun Kabir

Department of Management Accounting, Durban University of Technology, South Africa

Dr Md Humayun Kabir is an Associate Professor in the Department of Management Accounting at Durban University of Technology, South Africa. He received his MBA in Accounting and BBA Honours in Accounting from University of Chittagong, Bangladesh, and his Ph D in Business Management and Administration with specialization in Accounting from North-West University, South Africa. His research explores the impact of corporate management choice mostly around corporate social responsibility and corporate governance issues, but also around environmental management accounting, sustainability reporting, financial and non-financial disclosure, micro finance, corporate strategic planning, and privatization. Dr M H Kabir has published widely in leading corporate social responsibility, corporate governance, accounting, finance, and management journals.

Jyoti Devi (Brinda) Mahadeo

School of Business, Faculty of Commerce and Technology, Eastern Institute of Technology, Auckland Campus, New Zealand

Brinda is an Academic staff member in Management at the Eastern Institute of Technology. Prior to moving to New Zealand she has held a Senior Lectureship position at the University of Mauritius. She has been active in research for over a decade and her research is focused on Corporate Governance (CG), Corporate Social Responsibility and Entrepreneurship. She has won nationally funded research grants in the area of CG to investigate the implementation of the first code. Her current and future research is geared towards CG in Small and Medium Enterprises.

Stephen K. Nkundabanyanga

Department of Accounting, Makerere University Business School, Kampala, Uganda

Assoc. Prof. Dr. Stephen K. Nkundabanyanga, Ph.D, MBA, ACCA and B.COM, lectures at Makerere University Business School where he is also the Head of Department of Accounting. He received a Makerere University Busness School academic Excellence Award in 2015. Currently, he is also the Editor of the Makerere Business Journal. He has several years' work experience in the private sector including with Uganda Telecom and until March 2018, he was non-executive director at Top Finance Bank. Prof. Nkundabanyanga has authored several publications and is a member of numerous accounting and management associations. He is on editorial boards of refereed Journals such as *International Journal of Corporate Strategy and Social Responsibility, Journal of Public Budgeting and Financial Management* and *Journal of Accounting in Emerging Economies*.

Elewechi Okike

Department of Accounting, Covenant University, Ota, Ogun State, Nigeria

International Centre for Research in Accountability and Governance, Washington, UK

Elewechi Okike is Professor and CEO of the Academy for Excellence in Education (AFEXED), and Director of the International Centre for Research in Accountability and Governance (CENTRAG). She is a Visiting Professor at Covenant University, Ota, Nigeria, and has also held visiting professorial appointments at the University of Nigeria, Nsukka, and Qassim University, Saudi Arabia. She was Head of Accounting and Director of the research degree programme at Sunderland Business School. Her research interests are in international comparative studies in auditing, financial reporting and corporate governance. She promotes innovation and creativity in teaching accounting, auditing, corporate governance and the general areas of business and management, and has published extensively in these areas in highly rated peer-reviewed journals. She is on the International Editorial Board of *The Journal of Accounting and Auditing: Research and Practice* and is a consultant to the Institute of Corporate Governance of Nigeria. She was a Commonwealth Academic Staff Scholar, and represented Nigeria at the inaugural Conference of the Commonwealth Association for Corporate Governance (CACG), in Nairobi, Kenya, in 2000. She was in professional accounting practice before joining academia in 1980.

Peace Onuwabhagbe Okougbo

Department of Accounting, Covenant University, Canaan land Ota, Ogun State, Nigeria.

Peace Okougbo is a doctoral student, and an Assistant Lecturer at the Department of Accounting at Covenant University, Ogun State, Nigeria. She obtained first class honours degree and a Masters degree in accounting from Covenant University. Her research interests are in corporate governance, accounting education and accounting ethics. She won the best paper award for her paper titled 'Corporate Governance and earnings Management', at the Inaugural Conference of the African Accounting and Finance Association (AAFA) in Ghana, in 2011. She is actively engaged in research and has a number of publications in reputable peer-reviewed journals.

Semisi Prescott

Business Pathway, Unitec Institute of Technology. Auckland, New Zealand

Dr Semisi Prescott is a Senior Lecturer in Accounting and Finance at Unitec Institute of Technology. He currently heads the Postgraduate studies division within the business pathway working closely with both students and staff on projects related to the Pacific. His current research area is small and medium size enterprises in the South Pacific and business and culture. His PhD explored business sustainability based on Tongan experiences and has used this to expand on studying the drivers that represent the character of business development challenges in the South Pacific. His research has allowed him to be a member of the New Zealand government's social Services and Policy Evaluation Unit and as an advisor to different government ministries.

Juliet Wakaisuka-Isingoma

Department of Accounting, Makerere University Business School, Kampala, Uganda

Dr Juliet Wakaisuka-Isingoma obtained her PhD from the University of Nairobi, Kenya. She is a lecturer in the Faculty of Commerce, Department of Accounting, Makerere University Business School, Kampala, Uganda. She holds an MSc in Accounting and Finance, Bachelor of Commerce both from Makerere University, Uganda and a Diploma in Business Education from Kyambogo University, Uganda. She has over 12 years experience in the HE sector, and a number of publications to her credit, including "Corporate governance, firm characteristics, external environment and performance of financial institutions in

Uganda: A review of literature". She also has relevant industry experience.

Jimmy Walabyeki

Director, Lion Corporate Solutions, Kigali Rwanda

Jimmy is a Director of Lion Corporate Solutions, Rwanda, and a certified Corporate Governance Trainer with the Institute of Chartered Secretaries and Administrators (ICSA). He has written articles on Corporate Governance in Rwanda. His current research interests include Women on Boards of Directors, Bank Finance and its role in Corporate Governance in Companies, and Insider Trading in Capital Markets. He has trained members of Boards of Directors of both private and government companies, and has taught Corporate Governance and its applicability in Africa. He is a trained lawyer and a member of the East African Law Society. He completed his Master of Laws (LLM) degree at the University of Cape Town in December 2008. He was awarded the Rodman Ward Prize for the best essay in Corporate Governance at the University of Cape Town for the year 2008.

Corporate Governance in Commonwealth Countries: An Overview

Introduction

Corporate governance has ceased to be seen as the preserve of a few privileged countries with sophisticated stock exchanges. The need for good corporate governance has received global recognition, both at the level of national development and as part of the international financial architecture. The importance and the significance of corporate governance at the international level was encapsulated in the words of Jim Wolfensohn, former President of the World Bank, that "the proper governance of companies will become as crucial to the world economy as the proper governance of countries". Along these lines, the Commonwealth (herein after, CW) took a stride in using its unique position as a 'microcosm of the world' (Rt. Hon. Don McKinnon, former CW Secretary General) to capitalise on the benefits of globalisation. The process involved the recognition of the link between democracy, good governance and economic and social development. Thus, the stream of interest in corporate governance converged with increasing interest in associated issues of competitiveness, corporate citizenship, and social and environmental responsibility. Corporate governance has also been linked to the broader issues of national governance and action against corruption, with governments recognising that reform in the public sector would only be truly effective when accompanied by parallel reforms in the private sector. Therefore, this book, *Corporate Governance in Commonwealth Countries*, examines how the CW has achieved a degree of consensus in developing and promoting standards of corporate governance both in the public and the private sectors in member countries. It reveals the various organisations and institutions that have been at the forefront of supporting and promoting corporate governance in CW countries. However, before examining the contents of the book, the next section provides an overview of the Commonwealth, the establishment of the Commonwealth Association for Corporate Governance (CACG), and the approach adopted in developing and promoting good corporate governance in member countries.

An Overview of the Commonwealth

The Commonwealth brings together some 2.4 billion people of many faiths, races, languages, traditions and levels of development. It is one of the oldest groupings of the nations in the world, and represents one quarter of the world's population[1]. It emerged from the dismantling of the British Empire. In the decades since it rose from its 19th century Imperial roots, the Commonwealth has gone through significant changes, from an instrument of Imperial power in the early decades, to a voluntary association of independent countries. It was founded in 1931 to promote and achieve international understanding and world peace. These 53 member countries (see **Appendix 1**), mostly former colonies of the British Empire, span across Africa, Asia, the Americas, Europe and the Pacific. They range from the developed nations through to a diverse mix of emerging and transitional economies. They are amongst the world's smallest, largest, poorest and richest countries, and all members have an equal say irrespective of size and economic structure. The modern, multiracial CW emerged in 1949 when India became the first republican member. Today, 33 members of the CW are republics, 16 are constitutional monarchies, which recognize Queen Elizabeth II as their Head of State. And five have national monarchies of their own. The CW's growth accelerated in the 1950s and 1960s with the independence of many new members from Africa, the Americas, Asia, the Caribbean, the Mediterranean and the Pacific. They have developed beyond their roots, to form a unique platform for business growth on a global scale, and account for over 20 per cent of world trade. They consist of the type of networks that can promote trade and investment among member countries.

The developed countries within the CW, include the UK, Australia, Canada, New Zealand and Singapore. These countries have stable economies and democratic governments. Others like Malta, Malaysia and Barbados have relatively peaceful environments, comparatively moderate democracy and stable economy. But more than half of the developing countries, such as Lesotho, Pakistan, India, Mozambique amongst others, are politically unstable. Also there are many factors within these economies that affect their economic development. Help and support from other developed countries within the Commonwealth target extreme poverty and hunger, promote gender equality and develop global partnership between these countries.

All the CW countries subscribe to the CW's values and principles outlined in **The Commonwealth Charter**.

These principles centre around:

- democracy, the rule of law and good governance
- International peace and security, humans rights, gender equality and freedom of expression
- Sustainable development, protecting the environment and the role of civil society
- Access to Health, Education, food and shelter

They also share some commonalties, which minimise the 'cultural distance' and reduce transaction costs, thereby facilitating the flow of business and investment among member countries.

These include:

- Similar governmental structures and systems of parliament, civil service, public administration and law, rooted in the same tradition, and constantly being updated along similar lines, as CW governments share their experiences and methods in the world-wide programme of public sector reform.
- Similar systems of company law, commercial institutions, accounting, management and business practices;
- A common working language;
- An organisational and technical assistance structure which enables governments, the private sector and professional associations to regularly meet, debate and develop policies and ideas to promote a positive policy environment, including major events such as the Commonwealth Business Forum and the Commonwealth Heads of Government Meetings, amongst others.

The members are committed to ensuring that 'the Commonwealth is an effective association responsive to members' needs, and capable of addressing significant global challenges of the future'[2]. They aspire 'to a Commonwealth that is a strong and respected voice in the world, speaking out on major issues, that strengthens and enlarges its networks; that has a global relevance and profile; and that is devoted to improving the lives of peoples of the Commonwealth'[3].

Every two years CW leaders meet at the Commonwealth Heads of Government meeting (CHOGM) in a different CW country to review global political and economic development and to take strategic overview of the CW's working support of its member countries' interests. The most recent of such meetings was held in London in April 2018. Her majesty the Queen is the Head of the Commonwealth and attends the biennial Commonwealth Heads of Government Meeting, although it was announced at the CHOGM 2018 that Prince Charles will take over as the Head of the CW.

Background to the Establishment of the Commonwealth Association for Corporate Governance (CACG)

Whilst issues relating to corporate governance came into the limelight following the spate of corporate collapses in the UK in the mid-1980s, the publication of the Cadbury Code in the UK in 1992 gave it an added dimension. At the same time international interest in corporate governance increased with the financial crises in East Asia in the 1990s. These events drew attention to the vulnerability of cross border investments and the need for more effective governance of corporations, whether in the public or the private sector. This led to a number of world-wide initiatives for promoting corporate governance. Global standards of corporate governance are now recognised as a crucial part of the global financial architecture.

However, the particular interest in corporate governance in the Commonwealth emerging markets had its origins not so much in the context of the private sector financial systems as in the need to improve the performance and then to privatise state enterprises. The weaknesses of state enterprise boards, the issues of the accountability of the chief executive and the management team to the board and to their supervising ministry, the strategic effectiveness and the operational efficiency of state enterprises, had long been matters of concern in the CW. Many member countries experienced an increase in the privatisation of state enterprises, and consequently suffered from a 'post-privatisation policy vacuum'. This meant that the policy and the institutional environment were inadequate to cope with the demands of privatisation. Whilst the CW recognised, monitored and evaluated this lack of essential institutional fabric to sustain fair competition and appropriate regulation through its regular policy workshops, its view was that the scale of agenda required to deal with this policy vacuum was enormous. And, the time required for implementing appropriate policies, extensive. Hence, it concentrated on a few powerful instruments, which have significant short-term impact. This focus, according to the CW, was not at the expense of longer-term actions necessary to build the essential elements of an effective regulatory environment, but that it was the earliest phase of the longer and wider process of reform and capacity building for the 'post-privatisation environment'. Accordingly, the CW identified corporate governance as an important policy instrument requiring its technical assistance. Although corporate governance was not to be seen as some sort of panacea, nevertheless, the CW believed that the widespread practice of good corporate governance could help to achieve multiple objectives in emerging markets.

The other reason given by the CW for concentrating on corporate governance as a policy instrument was the relative ease with which capacities could be developed through institution building and training. It described this as a 'critical mass' that could be established (for example, Institute of Directors and Company Secretaries) in a matter of months rather than several years, while other policy reforms often took much longer.

The CW approach to corporate governance was an attempt to achieve the optimum combination of the fundamental imperatives of performance, conformance and consensus in the circumstances of a particular country. Therefore, the CW recognised corporate governance as the means, which, among others, would contribute to (Armstrong, 2000:59):

- Increased probity, efficiency and effectiveness of the financial markets, which in turn increases direct and portfolio investor confidence to commit long term funds to the country;
- Improved operational and strategic competitiveness of Commonwealth enterprises operating in the global market;
- The self-regulation of corporations, especially newly privatised utilities and public service enterprises, thereby facilitating more effective privatisation.

For these reasons, corporate governance was perceived as a powerful 'lever' for change, and included as a high priority item on the agenda. Thus, at the Commonwealth Business Forum in 1997 the recommendation was made that "*capacity should be established in all Commonwealth countries to create or reinforce institutions to promote good corporate governance; in particular, codes of good practice establishing standards of behaviour in the public and private sector should be agreed to secure greater transparency and to reduce corruption*". This resolution was endorsed in the Commonwealth Heads of Government Edinburgh Economic Declaration in October 1997.

The Commonwealth Association for Corporate Governance (CACG) was established in April 1998 in response to the Edinburgh Declaration, with membership drawn from various organisations, institutions and individuals within the Commonwealth. The CACG was a focal point for promoting corporate governance and encouraging national professional institutes in member countries to become part of this global network of international professional institutes. It had a five-year target in which it hoped to establish, through these professional institutes, self-sustaining institutional capacities for corporate governance (or branches of regional institutions among small countries) to cover each of the CW emerging markets, developing countries and associate states. The aim was, not only to promote and safeguard widespread practice of the highest standards of business direction in the private and public sectors, but also that these were directly linked to national priorities and concerns.

The CACG received financial support from the Commonwealth Secretariat, the World Bank, the Government of New Zealand, Commonwealth Fund for Technical Co-operation (CFTC) and other international agencies[4].

CACG Guidelines

In addition to the activities for promoting corporate governance, the CACG also issued Guidelines. The development of the CACG Guidelines, "*Principles of Corporate Governance in the Commonwealth*", and its adoption[5] in 1999 constituted an important aspect of the Commonwealth corporate governance initiatives. These Guidelines, which were intended to facilitate best business practices and behavior in the private or public sector, were neither mandatory nor prescriptive. They were also designed to be evolutionary in concept in order to be able to respond to further developments in corporate governance.

The CACG Guidelines comprise 15 principles of corporate governance (**Appendix 2**), which were structured to accommodate the needs of emerging and transitional economies, which make up a large part of the CW, as well as those of the more advanced economies and their international investors. The Guidelines were aimed primarily at boards of directors with a unitary board structure, most common in CW countries. Thus they applied to boards of directors (executive and non-executive) of all forms of business enterprises, including non-governmental organisations and agencies. The Guidelines were prepared because the work on national capacity building had demonstrated the need for an indicative code, which could be readily adapted by the national task forces to develop their own guidelines. Following the adoption of the main Guideline in 1999, other guidelines and codes have since been issued and adopted in many emerging economies within the CW. South Africa was the first African country in the CW to put in place appropriate corporate governance principles, and this was before the establishment of the CACG.

In 2000, the CW introduced an initiative, which focused on the banking sector, working through the various Central Banks. This was because:

- The banking sector is absolutely critical, and can be likened to the bloodstream of the economy – if the banking sector is healthy the rest of the economy can be strong; if the banking sector is poisoned by poor corporate governance the whole economy will be infected;
- In many developing countries the equity markets are small and do not play a strong role in the national capital markets, as many companies rely more on debt finance from their banks; in this context there are no institutional investors to perform the powerful role to encourage corporate governance as in OECD countries, but this function can be fulfilled to some extent by the banks;
- Central Banks can exert moral suasion and influence over the commercial banks and set requirements for all licensed commercial banks in accordance with the standards set by the Bank for International Settlements; the commercial banks can in turn recommend good corporate governance practices for their corporate customers (including majority of the private and family owned companies which are not publicly listed and subject to the stock exchange) in order to reduce their risk (and possibly enable improved interest rates).

Corruption in CW Countries

Corruption is a problem that has attracted much attention in recent times, especially following the recent debacles of many well-known companies, including Enron, Worldcome, Polly Peck amongst others. "Anti-corruption is crucial to corporate governance because bribery and corruption damage the governance of corporations and impede efforts to develop corporate integrity" (Luo, 2007:160). The CW is aware that if its corporate governance initiatives are to succeed, the issue of corruption cannot be overlooked. Corruption can take many forms and occurs within the public and the private sectors, and is not necessarily associated with any particular type of political/social system, form of government, or level of economic development. Bad governance and corruption are a hindrance to economic development. Accordingly, the CW is aware of this and is tackling the problem head-on. All CW countries who have not already done so are required to develop and implement their own national good governance and anti-corruption strategies. All the authors who have contributed chapters to *Corporate Governance in Commonwealth Countries* have alluded to the problem of corruption in their various countries, especially in the developing economies.

About the Book

This book, which is a compendium of writings with contributions from accomplished authors across the Commonwealth provides evidence of successful outcomes of the national capacity-building programmes and efforts at various levels to address corruption within the public and private sectors. It consists of 18 chapters, divided into six parts. Parts I to V focus on each of the groupings of the CW countries – *Africa, Asia, Caribbean and Americas, Europe* and the *Pacific*. Each part presents some background information about each of the countries represented in the specific region, such as the year each country joined the CW, the World Bank

Classification of the country, official language, currency in use, the population, corruption perception index score, and the ease of doing business in each country, among others. Part VI presents insights into corporate governance developments in selected CW countries – *Ghana* (Chapter 6), *Malawi* (Chapter 7), *Mauritius* (Chapter 8), *Nigeria* (Chapter 9), *Pakistan* (Chapter 10), *Rwanda* (Chapter 11), *Swaziland* (Chapter 12), *Tanzania* (Chapter 13), *Tonga* (Chapter 14), *Uganda* (Chapter 15), *United Kingdom* (Chapter 16) and *Lake Chad* (Chapter 17). It ends with *Concluding Remarks and Recommendations* (Chapter 18). The chapters present the relationship of each country with the CW and the corporate governance developments in each country. Chapter 17 examines corporate governance challenges in the management of Lake Chad, an Ancient Lake surrounded by four countries - Chad (on the east of the Lake), Niger (on the North Western side of the Lake), Nigeria (on the West side of the Lake) and Cameroon (on the South of the Lake). The latter two countries are member states of the Commonwealth. The Lake Chad Basin Commission was formed in the 1960s after many African countries gained their independence from Britain.

The book adds to our knowledge on corporate governance at the international level, especially within the Commonwealth, comprising a unique collection of nations – ranging from the developed economies through to vastly differing levels of emerging economies at varying stages of transition. Academics, researchers, business and finance students, investors and government agencies with an interest in the Commonwealth will find the book authoritative and insightful.

Elewechi Okike

REFERENCES

Armstrong, P. (2000) "Corporate Governance in Commonwealth Countries" in *The Commonwealth Trade & Investment Almanac 2000-2001*, 58-63.

Commonwealth Association for Corporate Governance (CACG) (1999) *CACG GUIDELINES: PRINCIPLES FOR CORPORATE GOVERNANCE IN THE COMMONWEALTH (Towards global competitivenenss and economic accountability).* http://www.ecseonline.com/PDF/CACG%20Guidelines%20-%20Principles%20for%20Corporate%20Governance%20in%20the%20Commonwealth.pdf (accessed 9/1/19)

Commonwealth Business Council (1999), *The Commonwealth Trade & Investment Almanac 1999*. London: Hanson Cooke Ltd.

Commonwealth Business Council (2000), *The Commonwealth Trade & Investment Almanac 2000-*

2001. London: Kensington Publications Ltd.

Luo, Y. (2007) Global Dimensions of Corporate Governance, Blackwell Publishing Ltd.

Appendix 1: The 53 Countries that make up the Commonwealth

Country	Date Joined+	World Bank Classification*	Population**
Antigua & Barbuda	1981	High Income	102,000
Australia	1931	High Income	24,599,000
Bahamas, The	1973	High Income	395,000
Bangladesh	1972	Lower-middle-income	164,670,000
Barbados	1966	High Income	286,000
Belize	1981	Upper-middle-income	375,000
Botswana	1966	Upper-middle-income	2,292,000
Brunei Darussalam	1984	High Income	429,000
Cameroon	1995	Lower-middle-income	24,054,000
Canada	1931	High Income	36,708,000
Cyprus	1961	High Income	1,180,000
Dominica	1978	Upper-middle-income	74,000
Fiji	1970++	Upper-middle-income	906,000
Gambia, The	1965^	Low income	2,101,000
Ghana	1957	Low-middle-income	28,834,000
Grenada	1974	Upper-middle-income	108,000
Guyana	1966	Upper-middle-income	778,000
India	1947	Lower-middle-income	1,339,180,000
Jamaica	1962	Upper-middle-income	2,890,000
Kenya	1963	Lower-middle-income	49,700,000

Kiribati	1979	Lower-middle-income	116,000
Lesotho	1966	Lower-middle-income	2,233,000
Malawi	1964	Low income	18,622,000
Malaysia	1957	Upper-middle-income	31,624,000
Malta	1964	High Income	465,000
Mauritius	1968	Upper-middle-income	1,265,000
Mozambique	1995	Low income	29,669,000
Namibia	1990	Upper-middle-income	2,534,000
Nauru	1968	Upper-middle-income	14,000
New Zealand	1931	High Income	4,794,000
Nigeria	1960^^	Lower-middle-income	190,886,000
Pakistan	1947^^^	Lower-middle-income	197,016,000
Papua New Guinea	1975	Lower-middle-income	8,219,000
Rwanda	2009	Low Income	12,208,000
Samoa	1970	Upper-middle-income	196,000
Seychelles	1976	High Income	96,000
Sierra Leone	1961	Low Income	7,557,000
Singapore	1965	High Income	5,612,000
Solomon Islands	1978	Lower-middle-income	611,000
South Africa+++	1931	Upper-middle-income	56,717,000
Sri Lanka	1948	Lower-middle-income	21,444,000
St Kitts & Nevis	1983	High Income	55,000
St Lucia	1979	Upper-middle-income	179,000

St Vincent & Grenadines	1979	Upper-middle-income	110,000
Swaziland^^^^	1968	Lower-middle-income	1,367,000
Tonga	1970	Upper-middle-income	108,700
Trinidad & Tobago	1962	High Income	1,369,000
Tuvalu	1978	Upper-middle-income	11,000
Uganda	1962	Low Income	42,863,000
UK	1931	High Income	66,022,000
United Rep. of Tanzania	1961	Low Income	57,310,000
Vanuatu	1980	Lower-middle-income	276,400
Zambia	1964	Lower-middle-income	17,094,000

*databank.worldbank.org/data/download/pop.xls (accessed 23/11/2018)
* https://blogs.worldbank.org/opendata/new-country-classifications-income-level-2 (accessed 23/11/2018)
**World Bank list of Economies (June 2018) (https://hupo.org/resources/Documents/World%20Bank%20list%20of%20economies%20(June%202018).pdf)
+thecommonwealth.org/our-member-countries
++ rejoined in 1997 after 10-year lapse)
^ Left 2013, rejoined in 2018
^^ suspended 1995-1999
^^^ left 1972; rejoined in 1989
+++ (Statute of Westminster; left in 1961; rejoined in 1994
^^^^Also known as the Kingdom of eSwatini

Threshold	**GNI/Capita (current US $)**
Low income	<$995
Lower middle income	$996 – $3,895
Upper middle income	$3,896 – $12,055
High income	>$12,056

APPENDIX 2 – Executive Summary of CACG Guidelines (PART ONE)

The board should:

- Principle 1 – exercise leadership, enterprise, integrity and judgement in directing the corporation so as to achieve continuing prosperity for the corporation and to act in the best interest of the business enterprise in a manner based on transparency, accountability and responsibility;
- Principle 2 – ensure that through a managed and effective process board appointments are made that provide a mix of proficient directors, each of whom is able to add value and to bring independent judgement to bear on the decision-making process;
- Principle 3 – determine the corporation's purpose and values, determine the strategy to achieve its purpose and to implement its values in order to ensure that it survives and thrives, and ensure that procedures and practices are in place that protect the corporation's assets and reputation;
- Principle 4 – monitor and evaluate the implementation of strategies, policies, management performance criteria and business plans;
- Principle 5 – ensure that the corporation complies with all relevant laws, regulations and codes of best practice;
- Principle 6 – ensure that the corporation communicates with shareholders and other stakeholders effectively;
- Principle 7 – serve the legitimate interests of the shareholders of the corporation and account to them fully;
- Principle 8 – identify the corporation's internal and external stakeholders and agree a policy, or policies, determining how the corporation should relate to them;
- Principle 9 – ensure that no one person or a block of persons has unfettered power and authority on the board which is, inter alia, usually reflected by separating the roles of the chief executive officer and Chairman, and by having a balance between executive and non-executive directors;
- Principle 10 – regularly review processes and procedures to ensure the effectiveness of its internal systems of control, so that its decision-making capability and the accuracy of its reporting and financial results are maintained at a high level at all times;

- Principle 11 – regularly assess its performance and effectiveness as a whole, and that of the individual directors, including the chief executive officer;
- Principle 12– appoint the chief executive officer and at least participate in the appointment of senior management, ensure the motivation and protection of intellectual capital intrinsic to the corporation, ensure there is adequate training in the corporation for management and employees, and a succession plan for senior management;
- Principle 13 – ensure that all technology and systems used in the corporation are adequate to properly run the business and for it to remain a meaningful competitor;
- Principle 14 – identify key risk areas and key performance indicators of the business enterprise and monitor these factors;
- Principle 15– ensure annually that the corporation will continue as a going concern for its next fiscal year.

(**SOURCE**: Commonwealth Association for Corporate Governance, November 1999)

Endnotes

---

[1] http://thecommonwealth.org/about-us (accessed 10/1/19)

[2] http://thecommonwealth.org/about-us (accessed 10/1/19)

[3] http://thecommonwealth.org/about-us (accessed 10/1/19)

[4] The CACG is currently not on the list of active Commonwealth organizations held at the Commonwealth Secretariat. It ceased to exist due to lack of funds and the demise of one of its founders, Michael Gillibrand.

[5] The Guidelines were adopted in the Durban Communiqué (paragraph 41) by the Commonwealth Heads of Government at its biennial meeting

in South Africa in November 1999, and were simultaneously endorsed by the Commonwealth Business Leaders Forum, representing some 40 countries and 640 delegates.

Part I - Africa

Chapter 1

Commonwealth African Member States

1.0 Introduction

Many countries in Africa were colonies of the British Empire, and after independence, most of them voluntarily joined the Commonwealth of Nations. Table 1:1 presents an analysis of the nineteen African member states of the Commonwealth (CW). It shows the year they joined the CW, their official languages, capital cities and currencies.

Table 1:1 – African Member States in the Commonwealth of Nations

Country	Year+	Capital City	Official Language	Currency
Botswana	1966	Gaborone	Setswana/English	Pula
Cameroon	1995	Yaounde***	French/English	CFA France (CFAfr)
Gambia, The	1965^	Banjul	English	Dalasi
Ghana	1957	Accra	English	Cedi (GH₵)
Kenya	1963	Nairobi	Kiswahili/English	Kenya Shilling
Lesotho	1966	Maseru	Sesotho/English	Loti (plural maloti) M
Malawi	1964	Lilongwe	English	Malawi Kwacha (MK)

Mauritius	1992	Port Louis	English	Mauritian Rupee (MRs)
Mozambique	1995	Maputo	Portuguese	Mozambique metical (MT)
Namibia	1990	Windhoek	English	Namibia dollar (N$)
Nigeria	1960^^	Abuja	English	Naira (N)
Rwanda	2009	Kigali	English plus 3 others++	Rwandan franc
Seychelles	1976	Victoria	Creole/English/French	Seychelles rupee (SRs)
Sierra Leone	1961	Freetown	English	Leone (Le)
South Africa	1931	Pretoria	11 Official languages+++	Rand ®
Swaziland**	1968	Mbabane	siSwati/English	Lilangeni^^^
Uganda	1962	Kampala	English/Kiswahili	Ugandan shilling (Ush)
UR. Tanzania	1971	Dodoma	Kiswahili/English	Tanzanian shilling (Tsh)
Zambia	1964	Lusaka	English	Kwacha

**(Kingdom of eSwatini)
*** Yaounde is the constitutional capital; Douala is the economic capital
+ http://thecommonwealth.org/our-member-countries/ (accessed 29/11/2018)
++ Rwanda has four official languages: Kinyarwanda, English, French, Swahili
+++ South Africa has eleven official languages – Afrikaans, English, Ndebele, Northern Sotho, Sotho, SiSwati, Tsonga,
Tswana, Venda, Xhosa and Zulu
^ Left in 2013 and rejoined in 2018
^^ Suspended between 1995 and 1999
^^^ Plural emalangeni

Of the nineteen African States belonging to the CW, 16 of them are republics, and have executive presidents; Lesotho and Swaziland are national monarchies, whilst Mauritius has a prime minister as the head of government.

1.1 Year of Joining the Commonwealth

Table 1:1 reveals that most of the African States joined the CW in the 1960s, with a few exceptions. Ghana joined in 1957, Cameroon, Mauritius, Mozambique and Namibia joined in the 1990s, whilst Rwanda joined in 2009. The Gambia left the Association in 2013 and re-joined in 2018. Two countries, Mozambique and Rwanda that were not former colonies also joined the Commonwealth of Nations in order to tap into the perceived benefits of belonging to the Association. South Africa was a founder member of the CW in 1931 after the country gained independence, but it ceased to be a member of the Association in 1961, because of its apartheid policies, which were contradictory to the fundamental values of the CW. However, the country re-joined the Association in 1994, following democratic elections, which saw the end of the apartheid regime. Nigeria was suspended between 1995 and 1999 because of its perceived human rights abuses[6], and government actions that were not aligned with the values of the CW. It was welcomed back after the return to civilian rule, following a democratic process.

1.2 Official Language of the CW African Member States

All of the African States in the Commonwealth, with the exception of Mozambique (understandably so) have English as an official language, being the language 'inherited' from their colonial past. Mozambique's official language is Portuguese, being a former Portuguese colony. Although English is spoken there, it is not an official language. In addition to English, some countries such as Botswana, Cameroon, Kenya, Lesotho, Rwanda, Seychelles, South Africa, Swaziland, Uganda and the United Republic of Tanzania have additional official languages. It is interesting to note that although Mauritius was firstly a French colony (between 1767 – 1810) before it became a British colony (1810 – 1968), it does not have French as an additional official language. English is regarded as the global language of the world[7], with recognised status in 51 countries. It is spoken as a first or second language in most of the CW. The report of the Inter-Governmental Group on Criteria for Commonwealth Membership states that English is the symbol of Commonwealth heritage and unity[8]. Mark Robson, Director of English and Exams at the British Council states that "English makes a significant contribution to sustainable global development. It eases trade between countries that do not share a common language"[9]. It is used as a language of convenience.

1.3 Currency of the CW African Member States

Under colonial rule, financial transactions and trading activities were conducted using the official currency of Great Britain, pounds, shillings and pence. After gaining their independence most of the countries developed their own currencies, although it is interesting to see that some of the East African countries (Kenya, Uganda and Tanzania) still retain some semblance of their colonial past, including the term 'Shillings' in their national currencies. Unlike other regions of the CW (such as the Pacific and the Americas) where currency substitution would appear to be more rampant, most African states seem to have developed and retained their indigenous currencies.

1.4 Hosting of CHOGM

As mentioned earlier, the Commonwealth heads of government (CHOGM) meet biennially to discuss various issues affecting member states, as well as any contentious or current issues needing to be addressed. The meetings are held in a different member state and chaired by the President or Prime Minister of the host nation, who becomes the 'Chair-in-Office' until the next meeting[10]. Over the years the focus of such meetings have changed, and since 1997, CHOGM now focuses on a theme, chosen by the host nation, which forms the basis of their deliberations. The first CHOGM was held in Singapore in 1971. Since then African member states have also played host to these meetings at various times, and an African (Chief Emeka Anyaoku, a Nigerian) held the position of Secretary-General from 1990 to 2000[11]. Furthermore, Nigeria hosted an extraordinary meeting in Lagos[12], in January 1966 and CHOGM in Abuja, in December 2003. The CHOGM was held in Lusaka, Zambia in 1979; Harare, Zimbabwe, in 1991, when the country was still a member of the Commonwealth. In 1999, CHOGM was held in Durban, South Africa.

1.5 Economy and Governance in African Commonwealth Member States

The World Bank classifies economies into four income groups[13] –

1. High Income
2. Upper-middle-income
3. Lower-middle-income
4. Low income

Table 1:2 – Economy and Governance in the CW African Member States

Country	Year	World Bank Class.^	Popu-lation	CPI *Score (out of 100)	Overall Governance Score (out of 100)	IIAG Trends (2008 – 2017)**	Ease of Doing Business (1–190)^^
Botswana	1966	Upper-middle-income	2,101,000	**61**	**68.5**	**-3.7**	**81**
Cameroon	1995	Lower-middle-income	24,054,000	**25**	**47.7**	**-0.3**	**163**

Gambia, The	1965^	Low income	2,101,000	**30**	**54.9**	**+3.0**	**146**
Ghana	1957	Low-middle-income	28,834,000	**40**	**68.1**	**+0.9**	**120**
Kenya	1963	Lower-middle-income	49,700,000	**28**	**59.8**	**+6.1**	**80**
Lesotho	1966	Lower-middle-income	2,233,000	**42**	**57.1**	**0.0**	**104**
Malawi	1964	Low-income	18,622,000	**31**	**55.8**	**-1.4**	**110**
Mauritius	1992	Upper-middle-income	1,265,000	**50**	**79.5**	**-0.7**	**25**
Mozambique	1995	Low income	29,669,000	**25**	**51.0**	**-3.0**	**138**
Namibia	1990	Upper-middle-income	2,534,000	**51**	**69.3**	**+3.3**	**106**
Nigeria	1960^^	Lower-middle-income	190,886,000	**27**	**47.9**	**+2.8**	**145**
Rwanda	2009	Low income	12,208,000	**55**	**64.3**	**+5.9**	**41**
Seychelles	1976	High income	96,000	**60**	**73.2**	**+4.0**	**95**
Sierra Leone	1961	Low income	7,557,000	**30**	**50.9**	**+0.5**	**160**
South Africa	1931	Upper-middle income	56,717,000	**43**	**68.0**	**-0.6**	**82**
Swaziland	1968	Lower-middle-income	1,367,000	**39**	**48.7**	**+1.4**	**112**
Uganda	1962	Low income	42,863,000	**26**	**55.0**	**+1.4**	**122**
UR. Tanzania	1971	Low income	57,310,000	**36**	**58.5**	**+1.3**	**137**
Zambia	1964	Lower-middle-income	17,094,000	**37**	**56.2**	**+1.8**	**85**

^ https://blogs.worldbank.org/opendata/new-country-classifications-income-level-2018-2019 (accessed 7/12/2018)

*Corruption Perception Index 2017 (transparency.org)(accessed 3/12/2018)
**The 2018 Ibrahim Index of African Governance – Overall Governance Score (https://mo.ibrahim.foundation/iiag/2018-key-findings/) (accessed 3/12/18). The 2018 IIAG is the 12th iteration. It has been produced annually since 2007, and has been refined and strengthened by the Board of the Foundation, the IIAG Advisory Council and others.
^^ Ease of Doing Business Index 2018 (1 – 190) (1= most business-friendly regulations) (World Bank data) (www.doingbusiness.org/content/dam/doingBusiness/media/Annual-Reports/English)

This is based on the Gross National Income (GNI), calculated in US dollars, using the *Atlas* Method.[14]

Table 1:2 provides the World Bank Classification of the CW African States. These classifications are not static but do change as circumstances within each particular country change. This means that a country can be classified as a lower-middle-income country in one year, and move into a different category in another year depending on factors such as income growth, inflation, exchange rates, and change in population. Whilst the World Bank provides financial assistance to most of these developing economies, this classification based on GNI does not influence their lending decision. The thresholds have moved down slightly from Low-income (<$1,005); Lower-middle-income ($1,006 – 3,955); Upper-middle-income ($3,956 – 12,235) and High-income (>12,235) because of low price inflation and the strengthening of the dollar[15].

The information on Table 1:2 shows that only one African CW member state, Seychelles, is classed as a high-income economy. Most of the other African economies fall into the lower income categories. Only four (Botswana, Mauritius, Namibia and South Africa) out of the 19 African member states are in the Upper middle-income category. Seven of the states (Cameroon, Ghana, Kenya, Lesotho, Nigeria, Swaziland and Zambia) belong to the lower middle-income category, whilst the remaining eight (The

Gambia, Malawi, Mozambique, Rwanda, Sierra Leone, Uganda and the United Republic of Tanzania) are low-income economies.

1.5.1 Size of the CW African member States

Interestingly, Seychelles is the smallest African country in the Commonwealth, in terms of the size of its population. This is not to suggest that the size of the country is responsible for its economic growth. This is far from the truth. In fact it is reported[16] that the small size of the country makes it vulnerable economically. It is an isolated island with limited natural resources and at some point was wholly dependent on tourism until that industry became competitive and the government made efforts to diversify the economy, encouraging farming, fishing and manufacturing in the whole country, in addition to privatizing some of the state-owned parastatals. Although the country also suffered from economic decline during the financial crisis, with the help of the IMF and the economic reforms initiated by the government, the economy bounced back, and has continued to experience growth and attracting foreign direct investment. This is why good governance and effective leadership is crucial for African economies whether or not they are members of the CW.

The population of the African CW member states, ranges from 96,000 (Seychelles) to 190,886,000 (Nigeria). The largest African member of the CW is Nigeria, with a population of over 190 million citizens. It is followed distantly by The United Republic of Tanzania, with a population of over 57 million. South Africa, with a population of over 56 million, is the third largest African country in the CW. Botswana, The Gambia, Lesotho and Namibia have populations of a little over 2 million, whilst Mauritius and Swaziland have populations of a little over 1 million. The other African member states fall in between.

1.5.2 Corruption in CW African Member States

Corruption as a concept is difficult to define and varies in context from one country to the other. It is generally viewed as the result of collective behaviour that has been institutionalized (Okike, 2004). Wallace and Parker (1991) identified three broad types of societies on the basis of the extent of corruption:

1) Non-corrupt – those that are relatively honest with little or no endemic corruption.
2) Highly corrupt – those with systematic (or pervasive) corruption with little or no control mechanisms in place.
3) Mildly corrupt – those with widespread corruption but with control mechanisms and sanctions against bad behavior.

Many African economies are underperforming because of corruption and poor governance of their economies, and yet they are mostly in need of foreign direct invest to help their economies grow. As Armstrong (2016:5) has rightly stated

> "corporate governance is strongly influenced by the level of sophistication of each economy and the accompanying development or maturity of its capital market ... In an era of the global economy, economic growth and development is driven as much by external factors as internal issues especially in relation to the flow of capital. The increasing significance of so-called emerging economies like China, India, Indonesia, Colombia, Brazil, Turkey and a number of others in the mainstream global economy in the 21st century has only sought to increase the competition for those capital flows."

Although this book is about corporate governance, the link between political, economic and social governance cannot be questioned. Good corporate governance may not prevail in an environment of poor public governance, and this is an issue confronting many developing economies, especially African states.

Table 1:2 shows the 2017 Corruption Perceptions Index (CPI) [17] for each of the African states in the CW, produced by Transparency International (TI), the global civil society organization leading the fight against corruption across the globe[18]. The CPI ranks countries and territories by their perceived levels of public sector corruption, on a scale of 0 to 100, where 0 is highly corrupt and 100 is very clean[19]. According to TI "this year's Corruption Perceptions Index highlights that majority of countries are making little or no progress in ending corruption"[20], the worst performing regions being Sub-Saharan Africa, with an average score of 32.

Table 1:2 reveals that of the nineteen African States in the CW, only five can be described as least corrupt, or mildly corrupt, following the classification of Wallace and Parker (1991). They are Botswana, with a CPI score of 61, Seychelles, 60, Rwanda, 55, Namibia, 51 and Mauritius, 50. The remaining African States have CPI scores of below 50, and fall into the classification of 'highly corrupt' countries. Interestingly, all the countries with CPI scores of 50 and above, with the exception of Rwanda, are in the High-income or Upper-middle-income category. Although Rwanda is currently classed as a Low-income economy, there is evidence (see Chapter 11) that the government has initiated a number of measures that are yielding positive results and ensuring rapid economic growth. TI reports that Sub-Saharan Africa with an average CPI score of 32 was the worst performing region, globally.

However, there is hope for Africa, through the activities of some African institutions, such as the African Union (AU), African Corporate Governance Network (ACGN) and the African Peer Review Mechanism (APRM), which are spear-heading some anti-corruption efforts in Africa. In 2018, the AU adopted the theme "*Winning the Fight against Corruption: A Sustainable Path to Africa's Transformation*". At the opening ceremony of the 30th Assembly of Heads of State and Government of the AU, which took place in Addis Ababa, Ethiopia, the current President of the Federal Republic of Nigeria, H.E Mr Muhammadu Buhari officially launched the theme on 28th January 2018[21]. The meeting specifically addressed the issue of tackling corruption in the different African States. In his address, President Buhari noted that, "Corruption is indeed one of the greatest evils of our time. Corruption rewards those who do not play by the rules and also creates a system of distortion and diversion thereby destroying all efforts at constructive, just and fair governance....It poses real threat to national security, unity and survival of the African State and people"[22].

TI observed from the CPIs that despite being the worst performing region as a whole, there are some African States such as Rwanda, Seychelles, Botswana and Namibia, amongst others, that are consistent in pushing back corruption, with notable progress[23]. These African States achieved higher CPIs than some Organisation for Economic Co-operation and Development (OECD) countries like Italy, Greece and Hungary[24].

TI notes that what distinguishes these top performing African States from the others is "political leadership that is consistently committed to anti-corruption. While the majority of countries already have anti-corruption laws and institutions in place, these leading countries go an extra step to ensure implementation"[25]. In Botswana, the least corrupt country in Africa, the government has a zero-tolerance approach to corruption[26]. These countries are a clear testament that corruption in Africa is manageable with well-sustained effort and good leadership.

1.5.3 Ibrahim Index of African Governance (IIAG)

The Mo Ibrahim Foundation defines governance as the provision of the political, social and economic public goods and services that every citizen has the right to expect from their state, and that a state has the responsibility to deliver to its citizens[27]. The Ibrahim Index of African Governance (IIAG) has been published since 2007 and was created to provide a quantifiable tool to measure and monitor governance performance in African countries, to assess their progress over time and to support the development of effective and responsive policy solutions. In the IIAG, country performance is measured across four key components that effectively provide indicators of a country's Overall Governance performance. These are:

- Safety and Rule of law
- Participation & Human Rights
- Sustainable Economic Opportunity
- Human Development

The purpose of the index is to help strengthen governance and leadership in Africa, which is the focus of the Mo Ibrahim Foundation (Hamdok, 2018).

Table1.2 presents the IIAG's (Overall Governance Score) (OGS) for the nineteen CW African member states. It shows that Mauritius has the highest score of 79.5, followed by Seychelles, with a score of 73.2. This would suggest that the governments in these states perform highly in relation to the four components noted above. Chapter 8, which examines corporate governance in Mauritius, shows the various steps taken by the government, which continues to give Mauritius the leading edge as the best performing economy in Africa. Table 1.2 shows that five states (Botswana, Ghana, Namibia, Rwanda and South Africa) have OGSs of over 60, whilst nine states (The Gambia, Kenya, Lesotho, Malawi, Mozambique, Sierra Leone, Uganda, Tanzania and Zambia) have OGSs of above 50. Cameroon, Nigeria and Swaziland have OGSs of 40 and above.

According to Hamdok (2018)[28], the IIAG is not a tool that intends to criticize governments: it is an objective exercise to help further the conversation on governance, to assess current and emerging trends, identify areas to improve, and to highlight and learn from success.

1.5.3.1 Ibrahim Index of African Governance Trends

Table 1.2 also shows the IIAG trends for the CW African member states for 10 years (2008 – 2017). The trends reveal that whilst a country like Botswana might appear to have OGS of 68.5, the country is actually experiencing a decrease of -3.7 in OGS over time. On the contrary, a country such as Kenya with a CPI score of 28, and an OGS of 59.8 shows the highest increase in OGS of +6.1 over the 10-year period. Overall, six countries (Botswana, Cameroon, Malawi, Mauritius, Mozambique and South Africa) show a decreasing trend in OGS over the 10-year period, whilst a country such as Lesotho remained consistent over the 10-year period. It showed neither increase nor decrease in OGS. The remaining twelve countries showed increasing trends over time. Rwanda and Seychelles are two countries that also showed high increasing trends of +5.9 and +4.0, respectively. Nigeria, which is often identified as one of the most corrupt countries in the world, with a CPI score of 27, has an OGS of 47.9 and an increase of +2.8 in the OGS over the 10-year period. This suggests that the efforts of the government to tackle corruption are having some positive outcomes.

The IIAG 2018 Report found that[29]

- African Governments have failed to translate economic growth into improved sustainable economic opportunity for their citizens.
- African governance progress is lagging behind demographic growth and youth expectations
- African countries show increasing divergence in overall governance performance
- Rule of Law, Transparency & Accountability are key to improving governance in Africa

The Report states that

> "since 2008 the African average score for Sustainable Economic Opportunity has increased by 0.1 point, an equivalent of 0.2%, despite a continental increase in GDP of nearly 40% over the same period. There has been virtually no progress in creating Sustainable Economic Opportunity, meaning it remains the Index's worst performing and slowest improving category.

It also stated, "there is also no strong relationship between the size of a country's economy and its performance in Sustainable Economic Opportunity".

The IIAG results confirm that Rule of Law and Transparency & Accountability are key pillars of good governance. It suggests that these two sub-categories show the strongest relationships with Overall Governance scores in Africa, with strong performance in these areas being the most common components of countries that perform well.

The Report also found that Transparency & Accountability are also strongly related to the Sustainable Economic Opportunity category and Business Environment sub-category, indicating that improvements in these areas will support progress and economic opportunity in Africa.

1.6 Ease of Doing Business

If CW member states are to attract foreign investments into their economies, not only must they be seen to promote probity, accountability and good governance, but also, they must remove bureaucratic bottlenecks in the business environment that will make it difficult to attract foreign direct investments. Along these lines, is the importance of recognizing the role that the private sector can play in enhancing economic growth and development. Appropriately, government policies, regulations, tax laws and other reforms must create an enabling environment for the private sector to develop, thrive and create job opportunities and help alleviate poverty. According to the World Bank Group President Jim Yong Kim[30], "sound and efficient business regulations are critical for entrepreneurship and a thriving private sector. Without them, we have no chance to end extreme poverty and boost shared prosperity around the world".

Since 2003, the World Bank has produced the *Doing Business* Index, which ranks economies based on the ease of doing business there, taking into account a number of factors, including trading regulations, property rights, contract enforcement, investment laws, the availability of credit and so on. Table 1:2 shows the index rankings of CW African member states in the 2018 *Doing Business* Report, which ranked the ease of doing business in 190 economies. The Table reveals that Mauritius and Rwanda have the best enabling environments for doing business amongst the CW African member states. Whilst it is understood that Mauritius has the best rank amongst the African member states, it is reported[31] that Rwanda is the biggest reformer in the history of Doing Business, and this is attributed to "strong leadership". Louise Kanyonga, Head of Rwanda Development Board's Strategy and Competiveness Department said[32] "in Rwanda, there was consensus in the years following the genocide and civil war that the private sector must be the key driver of economic transformation. The government set out to create an environment that would enable businesses and investments to thrive". Rwanda has become one of the world's fastest-growing economies, with a rapidly urbanizing middle class. According to Kanyonga, "private investments have grown by 60% over the last 8 years, and foreign direct investment alone has grown by close to 40%." Chapters 8 and 11 respectively provide further insights into government actions and reforms in both Mauritius and Rwanda that have seen these member states stand out as exemplars of economic growth, accountability and governance in CW African member states.

1.7 Conclusion

This overview of African CW member states shows that these states are at different levels of economic development. Those in the low-income category appear to have low CPI scores and also perform poorly on the IIAG. There is evidence that countries such as Mauritius, Seychelles, Rwanda, Namibia, Lesotho, among others are experiencing rapid economic growth and development and appear to have appropriate institutional structures in place to tackle corruption and similar vices that hinder economic growth. Hence they have IIAG scores that are reasonably high, even though the Report findings draw attention to shortcomings in some areas of governance. Nevertheless, the IIAG shows that African countries are showing an increasing trend in improvement in Overall Governance Score. Given this trend, better corporate governance becomes inevitable with proper checks and balances, and greater accountability and transparency. According to Kambi (2016:2) "corruption often flourishes where institutions are weak, gradually but surely eroding development and the very fabric of society".

REFERENCES

African Corporate Governance Network (ACGN) (2016), *State of Corporate Governance in Africa: An Overview of 13 Countries*. February

Fernandez, D. (2018) The 2018 Ibrahim Index of African Governance: reading the results,

(mo.ibrahim.foundation/news/2018/2018-ibrahim-index-african-governance-reading-the-results/) (accessed 12/12/18)

Hamdok, A. (2018 "The Ibrahim Index of African Governance: why it is important"
(mo.ibrahim.foundation/news/2018/Ibrahim-index-african-governance-important/) (accesed 12/12/18).

Kambi, S. (2016) *Foreword* to African Corporate Governance Network (ACGN) (2016), *State of Corporate Governance in Africa: An Overview of 13 Countries*. February

Kimbro, M.B. (2002), "A cross-country empirical investigation of corruption and its relationship to economic, cultural, and monitoring institutions: an examination of the role of accounting and financial statements quality", *Journal of Accounting, Auditing & Finance*, 17 (4): 325-49.

Mauro, P. (1995), "Corruption and growth," *Quarterly Journal of Economics*, 110 (3): 681-712.

Okike, E.N.M. (2004) "Management of crisis: the response of the auditing profession in Nigeria to the challenge to its legitimacy", *Accounting, Auditing & Accountability Journal*, 17 (5): 705-730.

Wallace, R.S.O. and Parker, R.H. (1991), "Corruption, Society and Auditing", *paper presented at the 3rd Annual Accounting, Business and Financial History Conference, Cardiff, 18-19 September*.

Endnotes

---

[6] Especially following the execution of Ken Saro-wiwa and eight others on the first day of CHOGM 1995. (https://en.m.wikipedia.org/wiki/Commonwealth_Heads_of_Government_Meeting) (accessed 12/12/18)

[7] https://englishlive.ef.com/blog/english-in-the-real-world/english-became-the-global-language (accessed 12/12/18)

[8] https://en.m.wikipedia.org/wiki/English_in_the_Commonwealth_of_Nations (accessed 12/12/18)

[9] Mark Robson, Director of English and Exams at the British Council. (https://www.britishcouncil.org/sites/default/files/english-effect-report-vz.pdf) (accessed 9/12/18)

[10] (https://en.m.wikipedia.org/wiki/Commonwealth_Heads_of_Government_Meeting) (accessed 12/12/18)

[11] http://thecommonwealth.org/our-member-countries/nigeria (accessed 3/12/18)

[12] https://en.wikipedia.org/wiki/Commowealth_Prime_Ministers%27_Conference (accessed 3/12/2018)

[13] The threshold used for classifying the economies is as follows:

Threshold	**GNI/Capita (current US $)**
Low income	<995
Lower middle income	996 – 3,895
Upper middle income	3,896 – 12,055
High income	>12,056

[14] According to the World Bank, the U.S dollars are converted from local currency using the *Atlas* method, which uses a three-year average of the official exchange rate adjusted for inflation, and does not account for differences in domestic price levels. The method lessens the effect of exchange rate fluctuations and abrupt changes. It is the method the Bank uses in determining development or measure welfare, the World Bank is of the view that it is a useful and easily available indicator that is closely

associated with other nonmonetary measures, such as the quality of life, life expectancy at birth, mortality rates of children, and enrollment rates in school. However, the limitations of the GNI include the fact that it may be underestimated in lower-income economies that have more informal subsistence and it does not reflect inequalities in income distribution (datahelpdesk.worldbank.org).

[15] https://blogs.worldbank.org/opendata/new-country-classifications-income-level-2018-2019 (accessed 7/12/2018)

[16] http://thecommonwealth.org/our-member-countries/seychelles/economy (accessed 4/12/18)

[17] The CPI was launched in 1995 and has been widely accredited with putting the issue of corruption on the international agenda.

[18] In 2009, TI published *Global Corruption Report,* which documented in unique detail the many corruption risks for businesses, ranging from small entrepreneurs in Sub-Saharan Africa to multinationals from Europe and North America. It also discussed the most promising tools to tackle corruption in business, and identified pressing areas for reform and outlined how companies, governments, investors, consumers and other stakeholders can contribute to raising corporate integrity and meeting the challenges that corruption poses to sustainable growth and development.

[19] https://www.transparency.org/news/feature/corruption_perception_index_2017 (accessed 5/12/2018)

[20] https://www.transparency.org/news/feature/corruption_perception_index_2017 (accessed 5/12/18)

[21] https://au.int/en/pressreleases/20180129/president-buhari-nigeria-launches-au-theme-year-2018-fighting-corruption (accessed 6/12/2018)

[22] https://au.int/en/pressreleases/20180129/president-buhari-nigeria-launches-au-theme-year-2018-fighting-corruption (accessed 6/12/2018)

[23] Regional Analysis by Paul Banoba

(https://www.transparency.org/news/feature/a_redefining_moment_for_africa) (accessed 6/12/2018)

[24] Regional Analysis by Paul Banoba

(https://www.transparency.org/news/feature/a_redefining_moment_for_africa) (accessed 6/12/2018)

[25] Regional Analysis by Paul Banoba

(https://www.transparency.org/news/feature/a_redefining_moment_for_africa) (accessed 6/12/2018).

[26] See https://neweralive.na/2018/02/07/how-president-khama-tackled-corruption/

[27] https://mo.ibrahim.foundation/iiag/ (accessed 9/12/2018)

[28] Dr Abdalla Hamdok is the Deputy Executive Secretary, Economic Commission for Africa (UNECA) and Chair of the Ibrahim Index of African Governance Advisory Council.

[29] https://mo.ibrahim.foundation/iiag/2018-key-findings/ (accessed 12/12/18).

[30] https://www.worldbank.org/en/news/immersive-story/2018/10/31/doing-business-2019-a year-of-record-reforms-rising-influence (accessed 14/12/18)

[31] https://www.worldbank.org/en/news/immersive-story/2018/10/31/doing-business-2019-a year-of-record-reforms-rising-influence (accessed 14/12/18)

[32] https://www.worldbank.org/en/news/immersive-story/2018/10/31/doing-business-2019-a year-of-record-reforms-rising-influence (accessed 14/12/18)

Part II - Asia

Chapter 2

Asian Countries in the Commonwealth

2.0 Introduction

Seven of the 53 countries in the Commonwealth are from Asia. The region has the highest number of CW citizens. Of the 2 billion citizens within the CW, more than 70 percent are from Asia. Table 2:1 provides a detailed analysis of the countries, the year they joined the CW, their capital cities, official languages and currencies.

2.1 Year of Joining the Commonwealth

India and Pakistan joined the CW after they gained their independence in 1947. Pakistan left the CW in 1972 and rejoined in 1989. Sri Lanka joined in 1948; Malaysia, 1957; Singapore in 1965 and lastly Brunei Darussalam in 1984. India was the first country with a republican form of government to join the CW, which paved the way for other countries, with adopted republican constitutions (or had a national monarch) to become CW members[33].

Table 2: 1 – Asian States in the Commonwealth of Nations

Country	Year	Capital	Official Language	Currency
Bangladesh	1972	Dhaka	Bangla	Taka (Tk)
Brunei Darussalam	1984	Bandar Seri Begawa	Malay	Brunci dollar (Br$)
India	1947	New Delhi	Hindi/English	rupee (Rs)
Malaysia	1957	Kuala Lumpur	Malay	ringgit**
Pakistan^	1947	Islamabad	Urdu	Pakistan rupee (Prs)
Singapore	1965	Singapore	English	Singapore dollar (S$)
Sri Lanka	1948	Colombo	Sinhala/Tamil	Sri Lanka rupee (SLRs)

**databank.worldbank.org/data/download/pop.xls (accessed 23/11/2018)
** ringgit or Malaysian dollar (M$)
+ English, Chinese (Mandarin), Malay, Tamil
^Joined 1947 (left in 1972, rejoined in 1989)

2.2 Official Language

Unlike the CW African member states, which mostly had English as their official language, only two countries, India and Singapore have English as their official language amongst the Asian member states. Brunei and Malaysia have Malay as their official language, Bangladesh, Bangla; Sri Lanka, Sinhala and Tamil; Pakistan, Urdu. India has Hindi as an official language in addition to English.

2.3 Currency of the CW Asian Member States

In terms of their currency, the Asian CW states have developed their own currencies, which have no semblance of their colonial past. Rather, the Indian Rupee is adopted in Pakistan and Sri Lanka, reflecting their close relationships with India.

Brunei and Singapore use their versions of the dollar. Both Brunei and Singapore have Currency Interchangeability Agreement between them. The Agreement was established in 1967 to promote monetary cooperation between them[34]. The Agreement reflects the strong bilateral relationship between the two countries, and financial cooperation between Autoriti Monetari Brunei Darussalam (AMBD) and the Monetary Authority of Singapore (MAS)[35].

2.4 Hosting of CHOGM

Like CW African member states, the Asian member states have also hosted the Commonwealth Heads of Government Meeting at various times. As a matter of fact, the first CHOGM was held in Singapore in 1971. India hosted the meeting in New Delhi in 1983, Malaysia in Kuala Lumpur in 1989, and Sri Lanka in 2013, and the fifth Secretary-General of the CW was an Indian, Mr Kamalesh Sharma. He held the position from 1 April 2008 to 31 March 2016.

2.5 Economy and Governance in Commonwealth Asian Member States

All economies desire to grow, whether in Africa or in Asia. However, corruption is a cankerworm that affects economies not only in Africa, but also in Asia, as well as other regions.

Table 2:2 provides additional information about the Asian CW member States, in relation to the World Bank classification of their economies, their population and their Corruption Perceptions Index scores, as produced by Transparency International, and the ease of doing business in Asia, produced by the World Bank.

Table 2: 2 – Economy and Corruption Perceptions in CW Asian Member States

Country	Year Joined	World Bank Classification*	Population	CPI Score (out of 100)	Ease of Doing Business Ranking (1 – 190)
Bangladesh	1972	Lower-middle-income	164,670,000	28	176
Brunei Darussalam	1984	High income	439,000	62	55
India	1947	Lower-middle-income	1,339,180,000	40	77
Malaysia	1957	Upper-middle-income	31,624,000	47	15
Pakistan	1947	Lower-middle-income	197,016,000	32	136
Singapore	1965	High income	5,612,000	84	2
Sri Lanka	1948	Lower-middle-income	21,444,000	38	100

*databank.worldbank.org/data/download/pop.xls (accessed 23/11/20180

The largest CW member state in Asia is India, with a population of over a billion citizens, followed by Pakistan, with a population of a little under 198 million citizens. Bangladesh, with a population of close to 165 million citizens, is the fourth largest member of the CW. Brunei Darussalam with a population under 450,000 is the smallest Asian country in the CW.

Two of the countries, Brunei Darussalam and Singapore are high-income economies; Malaysia is the only country in the upper-middle-income category, and the remaining four, Bangladesh, India, Pakistan and Sri Lanka are in the Lower-middle-income category. As mentioned earlier, these classifications are not static, but do change, as the circumstances of each country demand. The threshold used in the classification is provided in chapter 1.

2.5.1 Corruption in Commonwealth Asian Member States

The two countries (Brunei Darussalam and Singapore) with the lowest number of citizens and categorized as high-income economies have the highest CPI scores of 62 and 84, respectively. This suggests corruption is under control in these countries, and government policies aimed at curbing corruption seem to be effective. The CPI scores in the remaining five Asian member states are below fifty, suggesting there is some level of corruption existing in these countries. Going by these scores, Bangladesh is the most corrupt in this regional category. It scores 28, and ranks 142, out of the 180 countries. Pakistan scores 32, and ranks 117; India scores 40, and ranks 81 in the global ranking. TI's report of Asia Pacific countries suggests that their efforts at curbing corruption are slow and imperfect[36]. Mauro (1995) and Kimbro (2002) provide evidence that countries that grow at a very rapid pace tend to be more corrupt than those that grow moderately. This is probably due to the fact that a country can grow beyond the ability of its existing resources to cope with the growth. The World Bank notes that corruption impedes investment, with consequent effects on growth and jobs. Countries capable of confronting corruption use their human and financial resources more efficiently, attract more investment, and grow more rapidly[37]. According to TI, in some of the corrupt Asian countries, "journalists, activists, opposition leaders and even staff of law enforcement or watchdog agencies are threatened, and in most cases, even murdered"[38]. India is listed as one of the countries where such happens[39].

For corruption to be tackled effectively, these poor performing countries in Asia must have appropriate strategy to deal with the issue, including[40] putting in place laws and institutions that will prevent corruption from happening, reducing impunity for the corrupt, improving space for civil society to speak out and improving integrity and values. Moreover, the leadership of these countries must lead by example, too.

2.5.2 Ease of Doing Business in CW Asian Member States

Table 2:2 also shows the ease of doing business in Asian member states, (benchmarked to May 2018) according to the World Bank rankings of 190 countries. A high ease of doing business-ranking means the regulatory environment is more conducive to the starting and operation of a local firm[41]. The rankings reveal that Singapore has a more conducive regulatory environment for doing business than the other Asian member states, followed by Malaysia, with a world ranking of 15, and Brunei Darussalam with a world ranking of ranking of 55. With a world ranking of 176, it appears that Bangladesh has the most stringent regulatory environment for starting and operating a business, followed by Pakistan, with a world ranking of 136. Interestingly, these Asian member states with less conducive regulatory environments fall within the Lower-middle-income category, and also rank poorly on the TI's CPI score. This analysis indicates that these member states should do more to ease the regulatory burden of starting a business, and have effective measures in place to tackle corruption in order to attract foreign investments and move the economies into higher income categories.

TI notes that Asia Pacific states should not focus merely on their scores and rankings on the global scene, but much more, they should decide where to make substantial changes that will bring about real improvements in their countries. Their approach must be robust, otherwise there would be no improvement in the following year(s), or things could get even worse[42].

2.6 Asian Corporate Governance Association (ACGA)

Recognizing the importance of corporate governance to the long-term development of Asian economies and capital markets, the Asian Corporate Governance Association was founded in 1999, as an independent, non-profit membership organization dedicated to working with investors, companies and regulators in the implementation of effective corporate governance practices throughout Asia[43]. The establishment of ACGA became necessary given that poor corporate governance was identified as one of the root causes of the Asian crisis. The absence of effective discipline on corporate managers, coupled with complicated and non-transparent relationships between corporations, their owners and their finance providers, affected investors' confidence in the region's corporate sectors. Those economies that took early steps to improve corporate governance made quick recoveries from the crisis in comparison to those who did not address the issue. The Asian crisis was a pointer to the importance of good corporate governance not only for individual corporations to raise capital but also for an economy to achieve sustainable growth (OECD, 2001).

REFERENCES

Kimbro, M.B. (2002), "A cross-country empirical investigation of corruption and its relationship to economic, cultural, and monitoring institutions: an examination of the role of accounting and financial statements quality", *Journal of Accounting, Auditing & Finance*, 17 (4): 325-49.

Mauro, P. (1995), "Corruption and growth," *Quarterly Journal of Economics*, 110 (3): 681-712.

OECD (2001), *Corporate Governance in Asia: A Comparative Perspective*, Corporate Governance, OECD Publishing, Paris, https://doi.org/10.1787/9789264189300-en.

Endnotes

---

[33] http://www.commonwealthofnations.org/country/asia/ (accessed 3/12/18)

[34] www.mas.gov.sg/Currency-Interchangeability-Agreement-between-Brunei-Darussalam-and-Singapore.apx (accessed 15/12/18).

[35] www.mas.gov.sg/Currency-Interchangeability-Agreement-between-Brunei-Darussalam-and-Singapore.apx

[36] https://www.forbes.com/sites/ronakdesai/2018/03/07/india-continues-to-be-one-of-the-most-corrupt-countries-in-the-world/#564f98de79c6 (accessed 17/12/18).

[37] http://www.worldbank.org/en/topic/governance/brief/anti-corruption (accessed 17/12/18)

[38] https://www.transparency.org/news/feature/slow_imperfect_progress_across_asia (accessed 6/12/2018)

[39] https://www.transparency.org/news/feature/slow_imperfect_progress_across_asia (accessed 6/12/2018)

[40] https://www.transparency.org/news/feature/slow_imperfect_progress_across_asia (accessed 6/12/2018)

[41] www.doingbusiness.org/en/rankings (accessed 19/12/18)

[42] https://www.transparency.org/news/feature/slow_imperfect_progress_across_asia_pacific (accessed 17/12/18)

[43] https://www.acga-asia.org/who-we-are.php

Part III - Caribbean and Americas

Chapter 3

Caribbean and Americas Commonwealth Member States

3.0 Introduction

The thirteen CW member states from the Caribbean and Americas (C&A) are listed on Table 3:1. The Table shows the year each country joined the CW, their capital cities, official language and the currency in use. Ten of the thirteen states are islands. The exceptions are Canada, Belize and Guyana. Also, ten of the member states are constitutional monarchies that recognize Queen Elizabeth as their head of state, and have a prime minister as the head of government. A governor-general performs the ceremonial functions of the monarch. Dominica, Guyana, Trinidad & Tobago are republics. Whilst Guyana has an executive president, Dominica and Trinidad and Tobago have non-executive ceremonial presidencies, and an elected prime minister as head of government[44].

3.1 Year of Joining the Commonwealth

Canada was one of the founder members of the CW in 1931, when its independence was recognized under the Statute of Westminster[45]. The rest of the C&A states joined the CW in the 1960s (Barbados, Guyana, Jamaica and Trinidad & Tobago); the 1970s (The Bahamas, Dominica, Grenada, St Lucia and St Vincent & Grenadines) and the1980s (Antigua & Barbuda, Belize and St

Kitts and Nevis). All the member states in the Americas are multiparty democracies[46].

Table 3: 1 – Commonwealth Member States in the Caribbean and Americas

Country	Year Joined	Capital City	Official Language	Currency
Antigua & Barbuda	1981	St John's	English	Eastern Caribbean Dollar (EC$)
The Bahamas	1973	Nassau	English	Bahamian dollar (B$)
Barbados	1966	Bridgetown	English	Barbados dollar (Bds$)
Belize	1981	Belmopan	English	Belizean dollar (Bz$)
Canada	1931	Ottawa	English/French	Canadian dollar (C$)
Dominica	1978	Roseau	English	Eastern Caribbean dollar (EC$)
Grenada	1974	St George's	English	Eastern Caribbean dollar (EC$)
Guyana	1966	Georgetown	English	Guyana dollar (G$)
Jamaica	1962	Kingston	English	Jamaican dollar (J$)
St Kitts &	1983	Basseterre	English	Eastern Caribbean

Nevis				dollar (EC$)
St Lucia	1979	Castries	English	Eastern Caribbean dollar (EC$)
St Vincent & Grenadines	1979	Kingston	English	Eastern Caribbean dollar (EC$)
Trinidad & Tobago	1962	Port of Spain	English	Trinidad & Tobago dollar (TT$)

3.2 Official Language

Unlike the Asian States in the CW, which had only India using English as the official language, all the CW states in C&A use English as their official language. Perhaps this is not surprising given the fact that all these countries were colonies of Great Britain, and ten out of the thirteen countries still have the Queen as their ceremonial head of state. In addition to English, Canada also uses French as an official language. This is as a result of a Canadian law that came into force on September 9, 1969, which gives French and English equal status in the government of Canada, thus making them official languages[47].

3.3 The Currency of the Caribbean and Americas CW Member States

Given the colonial history of the C&A in the CW, it would be presumed that their currencies would have some semblance of the British Pounds and Shillings. However, this is not the case. Table 3:1 reveals that the dollar features prominently alongside the currencies of all the C&A states in the CW. Perhaps this is because the dollar is the most popular global currency accepted for trade throughout the world, as well as one of the most stable[48]. It makes up 64 per cent of all known central bank foreign exchange reserves[49].

3.4 Hosting of CHOGM

C&A States of the CW have hosted the bi-annual meeting of the CW Heads of Government at various times. The CHOGM was held in Ottawa, Canada in 1973; in Kingston Jamaica in 1975; Nassau, Bahamas in 1985; Vancouver, Canada in 1987, and Port of Spain in Trinidad and Tobago, in 2009. C&A countries also support and manage other CW initiatives. For example, the CW Association for Public Administration and Management, the CW of Learning and the CW Journalists Association have their head quarters in Canada, and the CW Library Association is based in Jamaica.

The first Secretary-General of the CW was Arnold Smith, a Canadian. He held the post from 1965 to 1975, followed by Sir Shridath Ramphal of Guyana until 1990[50].

3.5 Economy and Governance in the Caribbean and Americas CW Member States

Table 3:2 presents some economic data in relation to the C&A CW member states. The data suggest that there are no countries in the C&A that are in the low-income or low-middle-income category. All the countries fall within the upper-middle-income and high-income categories. In other words, the GNI per capita for this region did not fall below $3,896. This is an indication that the economies of these C&A CW States are reasonably healthy. However, Canada is the largest and only developed (or advanced) economy in this CW grouping. As mentioned in Chapter 1, these income groupings are not permanent but may change from year to year depending on the economic policies and productivity of each country.

3.5.1 Population of the Caribbean and Americas CW Member States

In relation to the population of the C&A CW member states, Canada has the highest population of 36,708,000, followed distantly by Jamaica with a population of 2,890,000, and Trinidad and Tobago, with a population of 1,369,000. The remaining ten C&A CW member states have population sizes of less than a million. Dominica and St Kitts and Nevis have population sizes of less than 100,000, implying that they are really small states.

Table 3: 2 – Economy and Corruption Perceptions in the Caribbean and Americas CW States

Country	Year Joined	World Bank Class*	Population	CPI Score (out of 100)	Ease of Doing Business (1 – 190)
Antigua & Barbuda	1981	High Income	102,000	N/A[51]	112
The Bahamas	1973	High Income	395,000	65	118
Barbados	1966	High Income	286,000	68	129
Belize	1981	Upper-middle-income	375,000	N/A	125
Canada	1931	High Income	36,708,000	82	22
Dominica	1978	Upper-middle-income	74,000	29	103

Grenada	1974	Upper-middle-income	108,000	52	147
Guyana	1966	Upper-middle-income	778,000	38	134
Jamaica	1962	Upper-middle-income	2,890,000	44	75
St Kitts & Nevis	1983	High Income	55,000	N/A	140
St Lucia	1979	Upper-middle-income	179,000	55	93
St Vincent & Grenadines	1979	Upper-middle-income	110,000	58	130
Trinidad & Tobago	1962	High Income	1,390,000	41	105

*databank.worldbank.org/data/download/pop.xls (accessed 23/11/2018)

3.5.1 Corruption Perceptions in Caribbean and Americas CW Member States

Table 3.2 also provides data on the CPI scores of the C&A CW States. It reveals that the most corrupt of the C&A CW member states is Dominica, scoring 29 out of 100 on the CPI, followed by Guyana, with a score of 38 out of 100. Jamaica and Trinidad and Tobago also score less than 50 on the CPI. This suggests that these two countries are highly corrupt. In a letter to the Editor of *Trinidad and Tobago NEWSDAY*[52], G.A Margues said

> "I have repeatedly said the country is corrupt to its core... Corruption has been in existence since independence, though well hidden at the top level of our society. Now it is generally expected and can be considered part and parcel of our culture. Every branch of Government is littered with the tentacles of corruption".

Similarly, in the *Jamaica Observer*[53], Greg Christie wrote:

> "Corruption in Jamaica is "entrenched and widespread". Jamaica must give serious consideration to what lies ahead should the Government and the country's lawmakers fail to decisively and aggressively confront its corruption problem".

Canada, with a CPI score of 82 is the least corrupt country in C&A, and ranks 8th globally.[54] The Bahamas and Barbados score 65 and 68, respectively. Grenada, St Lucia and St Vincent & Grenadines have CPI scores of above 50. In comparison with CW African member states and the CW Asian members states, the CW C&A member states have performed better in terms of their CPI scores and the status of their economies. However, Lipton, Werner and Gonçalves (2017) note that regional variation often mask the great deal of variation across countries. In Latin America and the Caribbean, corruption appears to be on par with other emerging market economies, but fare substantially worse than advanced economies. And there have been relatively small changes in efforts to get rid of corruption in Latin America, because once it is there, it is difficult to get rid of. Lipton et al (2017) suggest that countries need forceful policies that lead to changes in social perceptions so corruption is seen as the exception rather than the rule. And as corruption falls, governments will more easily detect those who remain corrupt because they will stand out. They also admit that it is not easy to change the behavior of people in a corrupt environment. It requires collective action, as isolated actions are unlikely to be effective. "A multifaceted and resolute push is needed to initiate positive dynamics out of the bad equilibrium. For that, strong leadership and society's support are key" (Lipton et.al. 2017).

3.5.1 Ease of Doing Business

Besides corruption, regulatory bottlenecks and unnecessary bureaucracy and administrative burden can limit a country's ability to attract foreign investments, or the ease with which to conduct business in the country. Appropriately, the World Bank ranks economies on their ease of doing business, from 1-190. A high ease of doing business-ranking means the regulatory environment is more conducive to the starting and operation of a local firm. The rankings are benchmarked to May 2018. Table 3.2 presents the rankings for the ease of doing business in the C&A member states. The data reveal that Canada, ranked 22 out of 190 economies, has the most conducive environment for doing business amongst the C&A member states. Jamaica follows distantly, with a ranking of 75, and St Lucia, with a ranking of 93. The remaining 10 C&A member states have rankings of over 100, implying that regulatory bottlenecks, administrative burdens and unnecessary bureaucracy constrain the ease with which business is conducted in these economies. Interestingly, most of these economies are in the Upper-middle-income classification, implying that their business environments are strong. Perhaps more could still be achieved with effective government policies to tackle corruption and ease regulatory bottlenecks that militate against attracting investments, innovation and creativity.

3.5.2 Corporate Governance Challenges in the Caribbean & Americas

In addition to corruption and a restrictive business environment, poor governance, whether in the public or private sector hinder economic growth. It has been established in academic literature that good corporate governance minimizes the incidence of corruption and improves the management of risks (Tricker, 2012). Sookram (2016) adds that good corporate governance improves access to capital and financial markets and attracts greater investment as well as strengthens management. Empirical research has revealed that over 84 percent of global institutional investors are willing to pay a premium for the shares of a well-governed company over one with a comparable financial record, but that is poorly governed (CIPE, 2008). Hence, the importance of having robust systems of corporate governance is hot on the agenda of all countries including the C&A, as they all scramble for their share of global finance and investment opportunities. But how companies are governed varies from country to country, depending on the effectiveness of the corporate governance standards and enforcement mechanisms in place. As would be expected, each of the CW C&A member state is implementing corporate governance systems, which reflect global best practices. However, Sookram (2016) proposes the adoption of responsible leadership approach to corporate governance in Caribbean member states because it is much more relevant to changing business contexts in the region. The role of responsible leadership in driving economic stability and tackling corruption is clearly evident in the case of Rwanda already discussed in Chapter 1. Member states in C&A can take a cue from that example.

Part VI presents corporate governance developments in selected CW member states, including Rwanda (Chapter 11).

As the most advanced economy in C&A, Canada has a robust system of corporate governance, derived from the British common law model although its close proximity to the United States means that it is sometimes influenced by developments in the US (MacDougall, et.al, 2018). Canada operates a principles-based model, in which listed companies have to comply with Canadian corporate governance principles (much similar to those of the UK), or explain why they have not. MacDougall et.al. (2018) state that institutional investors, the media and professional director associations play a key role in shaping corporate governance developments in Canada.

REFERENCES

Centre for International Private Enterprise (2008). *Corporate Governance for Emerging Markets*, July 31

Lipton, D., Werner, A. and Gonçalves, C (2017) 'Corruption in Latin America: Taking Stock,

https://blogs.imf.org/2017/09/21/corruption-in-latin-america-taking-stock/ (accessed 20/12/18).

MacDougall, A., Yalden, R. and Valley, J. (2018) 'Canada', *The Corporate Governance Review* (8th ed.), The Law Reviews, Law Business Research Ltd, London, UK.

Sookram, R. (2016) "Corporate Governance in the Emerging Economies of the Caribbean: Peculiarities, Challenges, and a Future Pathway", *The Journal of Values-Based Leadership*, 9(1) Article 9. Available at

http://scholar.valpo.edu/jvbl/vol9/iss1/9.

Tricker, B. (2012) *Corporate Governance: Principles, Policies and Practices* (2nd ed.). Oxford: Oxford University Press.

Endnotes

---

[44] http://www.commonwealthofnations.org/country/americas/ (accessed 6/12/2018)

[45] http://thecommonwealth.org/our-member-countries/canada (accessed 6/12/2018)

[46] http://www.commonwealthofnations.org/country/americas/ (accessed 6/12/2018)

[47] https://www.quora.com/Why-do-people-in-Canada-speak-French-as-well-as-English (accessed 6/12/2018)

[48] https://www.businessinsider.com/usd-countries-use-dollars-as-currency-2018-5?r=US&IR=T (accessed 20/12/2018)

[49] https://www.thebalance.com/world-currency-3305931 (accessed 7/12/2018)

[50] http://www.commonwealthofnations.org/country/americas/ (accessed 7/12/2018)

[51] CPI scores of Antigua & Barbuda, Belize and St Kitts and Nevis were not available on the world rankings provided by TI.

[52] Monday, 4 September 2017

[53] Sunday, 26 March 2017, Title of Article, "Jamaica's Future Choked by Cancer of Corruption"

[54] The least corrupt country globally, ranking 1st on the CPI score is New Zealand, scoring 89.

Part IV – Europe

Chapter 4

Commonwealth Member States in Europe

4.0 Introduction

There are only three European member states in the CW, all of which are island states and members of the European Union[55]. The United Kingdom (UK) was a founder member of the Commonwealth in 1931 when it recognised the independence of its dominions of Australia, Canada, New Zealand and South Africa under the Statute of Westminster[56]. After they attained independence and became republics, Cyprus and Malta joined the Commonwealth in 1961 and 1964, respectively. All three countries are multiparty democracies; in Cyprus the president is head of government and in Malta and the United Kingdom, the prime minister assumes the role[57].

Table 4: 1 – Commonwealth Member States in Europe

Country	Year Joined	Capital City	Official Language	Currency
Cyprus	1961	Nicosia	Greek/Turkish	Euro (€)
Malta	1964	Valletta	Maltese/English	Euro (€)
UK	1931	London	English	Pound sterling (£)

*Founder member

4.1 Hosting of Commonwealth Heads of Government Meeting

All three CW member states in Europe have at various times played host to the biennial CHOGM. In 1977, CHOGM was held in London, UK; in 1993, it was held in Limassol, Cyprus; in Edinburgh, UK in 1997 and Valletta, Malta, in 2005. The most recent CHOGM was again held in London in April 2018.

As the founding member of the CW, the UK hosts in London, a number of important divisions of the Association. This includes the Commonwealth Secretariat, Commonwealth Foundation, Commonwealth Business Council, Commonwealth Parliamentary Association, Commonwealth Local Government Forum, Association of Commonwealth Universities and Commonwealth Games Federation[58]. These are some of the various avenues through which CW member states collaborate to achieve their objectives.

4.2 Official Language of the European Member States of the CW

As most of the CW member states were former colonies of Britain, understandably many of these countries still retain English as their official language. Table 4:1 shows that Malta has English and Maltese as its official language, whilst the official language of Cyprus is Greek/Turkish. This notwithstanding, English is widely spoken in the south of Cyprus by the locals partially because of previous British rule and partially due to the tourism industry. In Northern Cyprus, Turkish was made the only official language by the 1983 constitution[59].

4.3 Currency of the CW European Member States

Cyprus and Malta are members of the European Union, and as would be expected, their currency is the Euro. UK did not sign up to the Euro, and so retained the British Pounds as its currency, even whilst still a member of the EU[60].

4.3 Economy and Governance in CW European Member States

Table 4:2 presents some data on the economy of the European member states of the CW as well as the CPI scores of the states. The table reveals that all three European member states are in the World Bank's high-income category.

Table 4: 2 – Economy and Corruption Perceptions in Commonwealth States in Europe

Country	Year Joined	World Bank Class*	Population	CPI Score (out of 100)	Ease of Doing Business (1 – 190)
Cyprus	1961	High Income	1,180,000	57	57
Malta	1964	High Income	465,000	56	84
UK	1931	High Income	66,022,000	82	9

The smallest CW member state in Europe in terms of population is Malta, with a population of less than 500,000. The UK has the highest population of over 66 million, and Cyprus has a population of a little over one million.

4.3.1 Corruption Perceptions in European CW Member States

All three European CW member states score reasonably highly on the CPI score. The UK has the highest CPI score of 82, the same with Canada (Table 3:2). Cyprus and Malta have scores of 57 and 56 respectively. It is noteworthy that some African member states such as Botswana and Seychelles have higher CPI scores of 61 and 60 respectively (Table 1:2), which are much higher than those of Cyprus and Malta. Also Singapore has a CPI score of 84 (Table 2:2), which is higher than that of Britain. Whilst it might be presumed that corruption would be much less amongst European member states, this analysis shows that some countries in Africa and Asia are taking more rigorous steps to tackle the problem of corruption, than some members of the European member states. As has been rightly inferred (Wallace and Parker, 1991; Kimbro, 2002) some countries are more corrupt than others, and it does not matter in what part of the globe they are based. It is the actions (or inactions) of governments that determine how corruption is perceived and addressed in any country.

4.3.2 Ease of Doing Business in European Member States

Table 4:2 also shows the ease of doing business in the three CW member states. Economies that rank highly (i.e. have a score of 1 to 20) appear to have less administrative bottlenecks for establishing businesses[61]. The table reveals that England has a world ranking of 9 out of 190 and implies that the economic environment is conducive for starting and growing a business.

Cyprus is ranked 57 among 190 economies in the ease of doing business. It averaged 45.27 from 2008 until 2018, and reached an all time high of 60 in 2014 and a record low of 26 in 2008[62]. Cyprus has a modern, free-market, service-based economy that gives international investors and domestic businesses confidence to invest, grow and prosper. The consumer market is strong and it has a high income per capita, generated mostly from its tertiary sector of services, which is considered the backbone of the economy. Yet the strong level of consumerism and a healthy national GDP has not impacted the cost of doing business[63]. It offers one of the lowest set up and operating costs within the EU[64]. Malta is ranked 84 out of 190 economies. The rank remained the same (84) in 2017, and averaged 80.60 from 2014 to 2018, reaching an all time high of 84 in 2017 and a record low of 76 in 2014[65]. The UK ranks 9 among the 190 economies in the ease of doing business. It averaged 7.27 from 2008 until 2018, reaching an all time high of 11 in 2012 and a record low of 5 in 2009[66]. According to the World Bank, the UK has one of the highest ease of doing business out of the countries of the world[67]. Forbes magazine reported that after the UK narrowly voted to leave the EU in 2016, the predictions were pessimistic that the UK economy would collapse, but surprisingly, the economy as a whole has remained relatively stable[68]. As a matter of fact, the present Business Secretary, Greg Clark boasted, "the UK is rightly recognized as having a leading business environment and responsible business practices, which are fundamental to making the UK one of the best places in the world to work, invest and do business".[69]

4.3.3 Corporate Governance in European Member States

The analysis on Table 4:2 suggests that the CW member states in Europe are less corrupt and have conducive environments for doing business. This implies they also have the propensity to attract foreign investments. Therefore, as would be expected these member states have corporate governance mechanisms in place for the efficient management of entities in which public funds are invested. Foreign investors are interested in the growth and security of their investments, hence the need to retain their confidence in the system. Appropriately, there is increasing pressure on boards of companies to take their fiduciary responsibilities more seriously. The various corporate scandals across the globe have helped to shape corporate governance developments and the UK would appear to be at the forefront of these developments. Given the colonial linkages with the UK, both Cyprus and Malta have also been influenced, not only by the UK model of corporate governance, but also by the developments of corporate governance in the UK. These developments are discussed in detail in Chapter 16.

REFERENCES

Kimbro, M.B (2002) "A cross-country empirical investigation of corruption and its relationship to economic, cultural, and monitoring institutions: an examination of the role of accounting and financial statements quality", *Journal of Accounting, Auditing & Finance,* 17(4): 325-349.

Smith, A. (1838) *The Wealth of Nations*, Ward Lock, London.

Wallace, R.S.O. and Parker, R.H. (1991) "Corruption, society, and auditing", *paper presented at the 3rd Annual Accounting, Business and Financial History Conference, Cardiff, 18-19 September.*

Endnotes

---

[55] As at the time of writing, the UK has voted to leave the EU, although the terms of leaving are yet to be decided.

[56] http://www.commonwealthofnations.org/country/europe/ (accessed 7/12/2018)

[57] http://www.commonwealthofnations.org/country/europe/ (accessed 7/12/2018)

[58] http://www.commonwealthofnations.org/country/europe/ (accessed 7/12/2018)

[59] https://en.wikipedia.org/wiki/Languages_of_C.yprus (accessed 21/12/18)

[60] At the time of writing, the UK has voted to leave the EU.

[61] https://tradingeconomics.com/malta (accessed 21/12/18)

[62] https://tradingeconomics.com/cyprus (accessed 21/12/18)

[63] https://www.tmf-group.com/en/news-insights/business-culture/top-challenges-cyprus/ (accessed 21/12/18)

[64] https://www.tmf-group.com/en/news-insights/business-culture/top-challenges-cyprus/ (accessed 21/12/18)

[65] https://tradingeconomics.com/malta (accessed 21/12/18)

[66] https://tradingeconomics.com//united-kingdom/ease-of-doing-business (accessed 21/12/18)

[67] https://tradingeconomics.com//united-kingdom/ease-of-doing-business (accessed 21/12/18)

[68] https://www.forbes.com/sites/kurtbadenhausen/2017/12/19/the-u-k-tops-forbes-best-countries-for-business-2018/#71c275dc26de (accessed 4/1/2019)

[69] https://www.frc.org.uk/news/december-2018-(1)/wates-principles-to-improve-corporate-governance-s (accessed 3/1/19)

Part V - Pacific

Chapter 5

Commonwealth Member States in the Pacific

5.0 Introduction

The eleven Pacific member states of the CW are listed on Table 5:1, which shows the year in which the member state joined the CW, the name of the capital city, the official language and the currency in use.

All the Pacific member states are discrete island nations, with the exception of Papua New Guinea.

5.1 Year Joined the CW

Table 5:1 shows that Australia and New Zealand joined the CW of Nations in 1931 as founding members, after their independence was recognised under the Statute of Westminster. Most of the other Pacific member states joined in the 1970s, after they gained independence, with the exception of Nauru that joined in 1968, and Vanuatu in 1980. Although Samoa joined in 1970, the country gained independence in 1962[70].

Five of the Pacific member states, Australia, New Zealand, Papua New Guinea, Solomon Islands and Tuvalu are constitutional monarchies, that recognise Queen Elizabeth II as head of state represented by a governor-general, although the prime minister is the elected head of government. Tonga is also a constitutional monarchy with extensive powers residing in the monarch, who also appoints the prime minister. Fiji[71], Samoa and Vanuatu are republics with a non-executive head of state and an elected prime minister as head of government. Kiribati and Nauru are republics with executive presidencies[72].

Table 5:1 – Pacific States in the Commonwealth of Nations

Country	Year Joined	Capital City	Official Language	Currency
Australia	1931	Canberra	English	Australian dollar (A$)
Fiji	1970	Suva	English	Fiji dollar (F$)
Kiribati	1979	Tarawa	English	Australian dollar (A$)
Nauru	1968	No capital	English	Australian dollar A$)
New Zealand	1931	Wellington	English/Maori	New Zealand dollar (NZ$)
Papua New Guinea	1975	Port Moresby	English	Kina (K)
Samoa	1970	Apia	Samoan	Tala or Samoan dollar (T)
Solomon Islands	1978	Honiara	English	Solomon Islands dollar (SI$)
Tonga	1970	Nuku'alofa	Tongan/English	pa'anga or Tongan dollar (TOP/T$)
Tuvalu	1978	Funafuti	Tuvalaun/English	Australian dollar/Tuvaluan dollar
Vanuatu	1980	Port Villa	Bislama/French/English	Vatu (Vt)

5.2 Official Language

English is the official language in all but one of the eleven Pacific member states of the CW. Samoa has Samoan, as its official language, although English is also spoken amongst the islanders. It is the second language, and is used for educational instruction[73]. Four of the Pacific member states have an additional official language. New Zealand has Maori, as a second official language, Tonga has Tongan alongside the English language; Tuvalu uses Tuvaluan in addition to English, and Vanuatu is an exception, with two additional languages, Bislama and French as official languages.

5.3 Currencies of the Pacific Member States

Table 5:1 presents the currencies of the Pacific member states. It reveals that in spite of their links with Britain, the currencies of the Pacific member states have no semblance with their colonial past. Rather, the dollar is the predominant currency in use, adapted to their individual countries. Australia uses the Australian dollar, so do three other states, Kiribati, Nauru and Tuvalu. New Zealand uses New Zealand Dollar. Samoa, Solomon Islands, Tonga and Tuvalu all use their versions of the dollar. The kina was introduced in Papua New Guinea in April 1975, and was used alongside the Australian dollar until 1 January 1976, when the dollar ceased to be a legal tender[74]. The currency in use in Vanuatu is Vatu. Thus Papua New Guinea and Vanuatu are the two member states that do not use any version of the dollar. Would these Pacific states be better off if they adopted the dollar instead? The Pacific states that issue their own currencies stabilize them against the US dollar because of their large and increasing trade with emerging Asia, which denominates its trade in US dollars. However, building on the theory of optimal currency peg, Freitag (2010) proffers his reason for 'dollarization' in the Pacific States. He argues that the replacement of their present currencies by the US dollar would strengthen the trade ties of these countries given that they already have trading relationships with East Asian countries, and this would reduce their transaction costs.

5.4 Hosting of CHOGM

Some Pacific member states have also played host to the biennial Commonwealth Heads of Government Meeting at various times. One of such meetings was held in Melbourne, Australia in 1981, Auckland, New Zealand in 1995, and Coolum, Australia in 2002.

A member of the Pacific region, Sir Donald McKinnon served as the Secretary-General of the CW from 2000 to 2008[75].

5.5 Economy and Governance in CW Pacific Member States

Table 5:2 shows the World Bank classification of the Pacific member states, their population and CPI scores provided by TI, as well as the ease of doing business in those states. The Table reveals that none of the member states falls within the low-income category. Australia and New Zealand are in the High-income category, Kiribati, Papua New Guinea and Vanuatu fall within the Lower-middle-income category, and the rest are in the Upper-middle-income category.

5.5.1 The Population Sizes of the Pacific Member States

The size of the population of these Pacific member states varies considerably, ranging from 11,000 to 24,599,000. Tuvalu has the smallest population of only 11,000, followed by Nauru, with a population of 14,000. Australia has the largest population in the region, with a population of a little less than 25 million. Papua New

Guinea has the next largest population of 8,219,000, followed by New Zealand, with a population of 4,794,000. The small island states spread across a unique diverse region made up of hundreds of islands, scattered over an area equivalent to 15 percent of the surface of the globe. Fiji is the largest member of this group of small island states.

Table 5:2 – Economy and Corruption Perception in the Pacific States of the CW

Country	Year Joined	World Bank Class	Population	CPI Score (out of 100)	Ease of Doing Business (1 to 190)
Australia	1931	High Income	24,599,000	77	18
Fiji	1970	Upper-middle-income	906,000	**	101
Kiribati	1979	Lower-middle-income	116,000	**	158
Nauru	1968	Upper-middle-income	14,000	**	*
New Zealand	1931	High Income	4,794,000	89	1
Papua New Guinea	1975	Lower-middle-income	8,219,000	29	108
Samoa	1970	Upper-middle-income	196,000	**	90
Solomon Islands	1978	Lower-middle-income	611,000	39	115
Tonga	1970	Upper-middle-income	108,700	**	91
Tuvalu	1978	Upper-middle-income	11,000	**	*
Vanuatu	1980	Lower-middle-income	276,000	43	94

*Ranking not available**These CPI scores are not available

Although there is great diversity within the group, these small states share similar challenges and opportunities given their size and remoteness. They are small narrowly based economies with limited natural resources and distantly away from major markets, and are vulnerable to external shocks. These factors, according to the World Bank can affect their growth as well as economic volatility[76]

5.5.2 Corruption Perception in the Pacific Member States

Transparency International states that corruption in the Pacific is a major impediment to good governance, and represents a significant risk to the stability of the South Pacific Region.[77] TI notes that in spite of relatively well-developed legal and policy frameworks, corruption is a major challenge in this region and manifests itself through various forms of bureaucratic, administrative and political corruption[78]. Table 5:2 reveals the CPI scores of the Pacific member states. The table shows that New Zealand, with a CPI score of 89, is the least corrupt country in the region, followed by Australia with a CPI score of 77. Papua New Guinea, with a CPI score of 29 is the most corrupt, followed by Solomon Islands, with a score of 39 and Vanuatu with a score of 43. The CPI scores for the remaining Pacific member states were not available as at the time of writing. However, Chêne (2010) notes that the Pacific Island countries are very diverse in terms of the extent of corruption risks and their respective level of governance, and law enforcement capacity. According to her,

> "In most of these countries, anti-corruption efforts are hampered by weak government capacity, low levels of state penetration, limited opportunities for public participation, as well as a lack of civil society involvement. Many of these issues are linked to factors associated with the size of the countries, the geographical features of the region and the challenge to integrate political institutions with strong cultural and tribal traditions that prevail in the region".

The best anti-corruption models suited to these small island states is still in doubt and is an interesting subject for further enquiry.

5.5.3 Ease of Doing Business

New Zealand is ranked the best place for doing business according to the World Bank's *Doing Business 2019* report. The report investigates regulations that enhance or constrain business activity in 190 countries. Since its first release in 2003, these series of annual reports have inspired reforms in many economies across the different regions. Table 5:2 shows the ease of doing business ranking in the rest of the Pacific member states. Australia ranks 18th in the World Bank ranking, but second best place to do business amongst the Pacific member states. It appears the small island states with corruption challenges according to TI, and ease of business ranging from a ranking of 90 for Samoa and 158 for Nauru, are facing challenges of attracting foreign investments. In a keynote address Batley (2007)[79] rightly pointed out that the rates of economic growth among Pacific Island countries have, with few exceptions, not kept pace with population increases so that real economic growth has essentially stagnated over a generation. He noted that economic growth in the Pacific countries has consistently underperformed that of other developing countries. His advice was that countries in the region needed to take action to attract long term investment, and one of the keys to attracting investment to particular markets is effective corporate governance – both in the sense of an effective regulatory environment, as well as individual corporate practice.

5.5.4 Accountability and Governance in the Pacific Member States

The Transparency International CPI scores and the World Bank Ease of Doing Business rankings are pointers to economies that are well managed and that have effective regulations for ethical business practices. Such economies are at the forefront of attracting investments across the global capital market. In a keynote address Batley (2007) said “in today’s globalized market, investors are looking for a predictable legal and regulatory environment, an effective resolution system, proven transparency of transactions, and evidence of good corporate governance. Individual enterprises – wherever they operate around the globe – are assessed on whether or not they have sound or poor corporate governance practices”.[80] Also, the adoption of robust corporate governance practices, which provide accountability to shareholders and transparency for stakeholders, has an important ‘signalling’ effect in the market, improving investor confidence and assisting companies to compete for scarce capital (Williams, 2016).

Appropriately most advanced economies like the UK, Australia and New Zealand keep on top of their governance game with robust corporate governance systems.

5.5.4.1 – Corporate Governance in Australia

Williams (2016) states that within Australia, there is a prevailing view that ‘good’ systems of corporate governance can support and improve company performance and this has influenced the extension of corporate governance beyond merely being a matter of legal compliance. Instead, the Australian Corporate Governance framework includes a mix of prescriptive and voluntary elements, including:

- **‘hard’ law**, being legally binding case law and legislative requirements, such as the *Corporations Act 2001* (Cth) (**Corporations Act**).
- **‘soft’ law**, being the listing rules of Australian Securities Exchange Limited (**ASX**) which principally have effect as a contract under law, and

- **non-binding guidelines**, most notably including the third edition of the ASX Corporate Governance Council's Principles and Recommendations (**ASX Principles**).

But the key features of corporate governance in Australia, according to Williams (2016), include:

- extensive regulation and personal liability of directors,
- a move towards principles-based systems of governance, and
- influential large institutional investors and a strong investment 'infrastructure'.

According to Williams, the aims of the corporate governance reforms in Australia have been:

- the development of systems of governance which will protect and support the sustainable growth of shareholder wealth, and
- increased disclosure of companies' activities, performance, position and prospects, supported by enhanced dialogue and interaction between boards, management and interested investors.

5.5.4.2 – Corporate Governance in New Zealand

In New Zealand, the Financial Markets Authority (FMA) has issued *Principles and guidelines* (A handbook for directors, executives and advisers) of non-listed and public sector entities. Recognizing that many of these entities have a significant impact on New Zealand's financial markets, and that a 'one size-fits-all' approach might not be appropriate, the purpose of the handbook is to assist directors, executives and advisers of non-listed and public sector companies, and other entities, to apply corporate governance principles to their particular entity. The principles are not legally binding, and reporting against them is voluntary. However, FMA (2018) states that the principles set out standards for corporate governance that directors and executives should apply, and report on, to their investors, shareholders and stakeholders.

Prior to this, the New Zealand Stock Exchange issued the NZX Corporate Governance Code 2017 for listed companies. The new updated NZX Code is the primary guidance practices for NZX listed companies and is structured around eight principles:

1. **Code of ethical behaviour** – directors should set high standards of ethical behaviour, model this behaviour and hold management accountable for those standards being followed throughout the organization.

2. **Board composition and performance** – to ensure an effective board, there should be a balance of independence, skills, knowledge, experience and perspectives.

3. **Board committees** – the Board should use committees where this will enhance its effectiveness in key areas, while still retaining board responsibility

4. **Reporting and disclosures** – the Board should demand integrity in financial and non-financial reporting and in the timeliness and balance of corporate disclosures.

5. **Remuneration** – the remuneration of directors and executives should be transparent, fair and reasonable.

6. **Risk management** – directors should have sound understanding of the material risks faced by the issuer and how to manage them. The Board should regularly verify that the issuer has appropriate processes that identify and manage potential and material risks.

7. **Auditors** – the Board should ensure the quality and independence of the external audit process.
8. **Shareholder rights and relations** – the Board should respect the rights of shareholders and foster relationships with shareholders that encourage them to engage with the issuer.

Listed companies are required to 'comply' with the recommendations of best practices under each principle or 'explain' why they have not complied; hence the NZX Code adopts a 'comply or explain' approach. This approach allows listed companies to apply other corporate governance practices that are more suited to the size and type of their business and explain this in their report.

The new NZX Code was issued following feedback received from two rounds of consultations with a wide range of relevant market participants, which commenced in 2015, to ensure that the views of all parties were taken into consideration. The Financial Markets Authority has approved the updated Code, which seeks to more closely align with the FMA's Corporate Governance in New Zealand Handbook - Principles and Guidance and the ASX Corporate Governance Principles and Recommendations.

5.5.4.3 – Corporate Governance in other Pacific member states

As member states of the CW that are more developed, New Zealand and Australia play a big role in supporting the smaller states in the Pacific region to help them grow and become sustainable economies. Although majority of the businesses in these small states are family-owned, many are seeing significant growth, hence are developing a greater appetite for good corporate governance. Good corporate governance will not only help them improve their performance, but it will also help them attract much needed finance as well as sustained economic growth.

The Australian government recently supported the International Finance Corporation (IFC), a member of the World Bank Group, to establish the Pacific Corporate Governance Institute (PCGI) with the aim of promoting good governance practices in the Pacific business sector. PCGI operates as an independent non-profit organization, which is promoting principles of good corporate governance to companies in Fiji, and operates as a centre of excellence for corporate governance in the broader Pacific region.[81] The IFC's Corporate Governance Program in the Pacific helps companies in the region adopt effective governance standards and practices 'by raising market awareness, building the capacity of regulators and local partners, and providing firm-level advisory services'[82]. Earlier, in 2007, the Australian government provided $100,000 to the Fiji Capital Markets Development Authority (CMDA) to enable it to establish its Corporate Governance Council, and support its efforts to promote good corporate governance practices[83].

The Pacific small island states receive support not only from the Australian and New Zealand governments, but also from various other organisations such as the Asian Development Bank[84], the OECD[85], Transparency International[86] and the World Bank, amongst others. Chapter 14 examines corporate governance in one of the small island states, Tonga, and provides some insight into the types of support provided to these small island states to develop good practices in corporate governance and the effectiveness of these support mechanisms.

REFERENCES

Batley, J. (2007) "Effective Governance: Attracting Long term Investment", being keynote address of the Australian High

Commissioner at the CMDA Conference, Radisson Resort, Denarau, 28 September

Chêne, M. (2010), Corruption Challenges in Small Island Developing States in the Pacific Region, U4 Anti-Corruption Resource Centre, Chr.Michelsen Institute (CMI), Transparency International

Financial Markets Authority (FMA) (2018) Corporate Governance in New Zealand: Principles and guidelines. 180228 Corporate Governance Handbook.

Freitag, S. (2010) "Choosing an Anchor Currency for the Pacific", Center for European Governance and Economic Development Research Discussion Papers No.112 October.

New Zealand Stock Exchange (NZX)

Williams, I. (2016) "Corporate Governance in Australia: A Snapshot", accessed at

https://herbertsmithfreehills.com/latest-thinking/corporate-governance-in-australia-a-snapshot.

World Bank (2019) Doing Business 2019, 16th Edition, International Bank for Reconstruction and Development, Washington.

Endnotes

---

[70] http://www.commonwealthofnations.org/country/pacific/ (accessed 7/12/2018)

[71] General elections were held in Fiji on 14 November 2018. The result was a victory for the ruling FijiFirst party of Prime Minister Frank Bainimarama, which received just over 50% of the vote and 27 of the 51 seats in Parliament, a loss of five seats.

[72] http://www.commonwealthofnations.org/country/pacific/ (accessed 8/12/2018)

[73] https://www.studycountry.com/guide/WS-language.htm (accessed 23/12/18)

[74] https://en.m.wikipedia.org/wiki/Papua_New_Guinea_kina (accessed/12/18)

[75] http://www.commonwealthofnations.org/country/pacific/ (accessed 8/12/2018)

[76] https://www.worldbank.org/en/country/pacificislands/overview (accessed 1/4/19)

[77] https://www.transparency.org.nz/pacific-region-transparency/ (accessed 4/1/19)

[78] https://www.u4.no/publications/corruption-challenges-in-small-island-developing-states-in-the-pacific-region (accessed 4/7/19.

[79] https://fiji.embassy.gov.au/suva/governance.html (accessed 4/1/2019)

[80] https://fiji.embassy.gov.au/suva/governance.html (accessed 4/1/2019)

[81] https://www.vilatimes.com/2018/03/07/new-zealand-is-not-trying-to-compete-with-china-in-the-pacific--nzs-prime-minister/ (accessed 4/1/19)

[82] https://www.vilatimes.com/2018/03/07/pacific-corporate-governance-institute-in-fiji-is-now-open/ (accessed 4/1/19).

[83] https://fiji.highcommission.gov.au/suva/cmdarelease.html (accessed 7/1-19)

[84] https://www.adb.org/sites/default/files/evaluation-document/36180/files/in17-14.pdf (accessed 7/1/19)

[85] https://www.oecd.org/derec/49414196.pdf (accessed 7/1/19)

[86] https://www.u4.no/publications/corruption-challenges-in-small-island-developing-states-in-the-pacific-region (accesed 7/1/19)

Part VI – Corporate Governance in Selected Commonwealth Countries

Chapter 6

Corporate Governance in Ghana

Dr Ibrahim Bedi

6.0 Introduction

British rule in Ghana ended on the 6th of March 1957, when the country gained independence. However, it was not until the 1st of July 1960 that Ghana became an independent Republic. By implication, therefore, Ghanaian laws and codes are largely British adopted and influenced. The Companies Act, 1960 (Act 179) as well as several other laws are replicas of the British law sometimes with moderate local contents. Ghanaian laws and codes are significantly and positively influenced and amended to reflect British legal architecture. In addition, the wave of corporate governance practices in the Commonwealth of Nations also contributed to influence Ghana to adopt several of the international codes, rules and principles of corporate governance.

Ghana has adopted international best practices on corporate governance promulgated by bodies such as the United Nations Conference on Trade and Development (UNCTAD), Organisation for Economic Co-operation and Development (OECD), Committee of Sponsoring Organizations (COSO), International Finance Corporation (IFC), World Bank and the Securities and Exchange Commissions. The United Nations Conference on Trade and Development's 'guidance on good practices in corporate governance disclosure' (2006) partly influenced Ghana's Securities and Exchange Commission (SEC) to develop a documented corporate governance code for financial market players. This was meant to promote corporate governance practices in line with global trends following the aftermath of the collapse of Enron, WorldCom and Arthur Anderson. This development paved the way for non-listed companies to rethink their corporate governance mechanisms.

Prior to SEC Code, the Companies Act, 1960 (Act 179) of Ghana offered limited guidelines for governance. Act 179 deals with the appointment of directors, relationship with shareholders, external audits, annual general meetings and preparation of annual reports. Multinational companies such as Barclays Bank, Standard Chartered Bank, British Tobacco and Unilever Ghana Limited were the known, respected and model companies for corporate governance. However, Unilever for example, had a duality of Chief Executive Officer and Board Chair.

The joint initiative of the World Bank and the International Federation of Accountants (IFAC) in 2004 on the assessment of accounting and auditing practices revealed certain systemic weaknesses and lack of enforcement in Ghana. The Report on Observance of Standards and Codes (2004:1) resulting from the assessment states:

> *"The policy recommendations provided in this report focus on improving statutory framework, strengthening enforcement mechanisms, upgrading professional education and training, and enhancing capacity of regulatory and professional bodies. A major recommendation is that an independent oversight body—Financial Reporting Council—should be established. This is in line with the global trend towards accountancy bodies being more open in their* ***governance*** *and regulatory processes and for more independent oversight or regulation in areas of public interest".*

The Institute of Chartered Accountants (Ghana) with financial and technical support of IFAC has revised its governance structures in line with the recommendations of the assessment. The Institute has included a paper in its professional examinations titled 'Governance, Ethics and Strategy' to promote corporate governance in Ghana. In addition, several fora and continuing development programmes have been held to promote governance among members of the Institute and the general public. However, Ghana has not established the Financial Reporting Council (FRC) due to legal and political reasons. The establishment of the FRC would eliminate or reduce the powers and supervisory role of the Institute of Chartered Accountants (Ghana) as well as the financial reporting supervisory roles of bodies such as the Bank of Ghana, the Ghana Stock Exchange and Ghana Insurance Commission. The presence and influence of these bodies in the financial reporting landscape has slowed the process for establishing the FRC in Ghana.

6.1 Changes in Corporate Governance Post-2006

The changes in corporate governance after 2006 can be grouped into two, Private Sector Governance and Public Sector Governance Changes.

6.1.1 Private Sector Governance Changes

The Securities and Exchange Commission issued the 'corporate governance guidelines on best practices' 2010 (commonly called the Corporate Governance Code 2010). SEC has oversight of broker dealers (20), custodians (10 banks), depositories (1), fund managers (120), investment advisory (6), issuing houses (1), mutual funds (30), primary dealers (17), registrars (4), trustees (5) and unit trusts (19).

The Code provides guidelines, which among others include the following:

i. Principles of Corporate Governance:
   Corporate governance must cover the rights of shareholders; equitable treatment of shareholders; role of stakeholders; disclosure and transparency; and the responsibilities of the board.
ii. Role of the Board
iii. Role of the Chairman
iv. Orientation and Training
v. Size of the Board:
   The Code suggests an appropriate size to be determined by the company but recommends 8 to 15 members
vi. Appointment of Board Members
vii. Multiple Board Seats:
   The Code requires that other directorships of each board member must be assessed before appointment so that board members who can be available or active are appointed.
viii. Separation of Board Chairman and Chief Executive:
   The Code requires separation of the Chairman and the Chief Executive Officer. However, the Code accommodates instances where one person can serve both roles. The approval of the SEC is required for such duality of role.
ix. Composition of the Board:
   The Code requires an appropriate mix of executive and non-executive members.
x. Term and Age Limits:
   The Code recommends at least a 3-year tenure for board members
xi. Access to Information
xii. Board Committee Meetings
xiii. Audit Committee and Other General Committees

xiv. The Code recommends appropriate committees to be formed but suggests at the least the formation of the audit and remuneration committees.

xv. Financial Affairs and Auditing

xvi. Audits and Rotation:
Internal and external audits are required as an important component of corporate governance. Rotation of key audit personnel is required to ensure independence.

xvii. Disclosures in Annual Reports:
The Code requires an explicit statement that a company has complied with its requirements.

xviii. Requirement of a Code of Ethics:
The Code requires companies to develop Code of Ethics to guide governance and operations.

Currently, SEC had drafted a new regulation on doing business for market operators which provides further guidance on governance called the 'Securities Industry (Market Operators) Conduct of Business Regulation, 2017.

The governance issues in the Regulation among others include:

i. The role and responsibilities of the Board

ii. The composition of the Board:
The Regulation requires three directors and at least, one-third shall be independent non-executive directors. New appointments to the Board cannot be made without the prior approval of the SEC.

iii. The operation of the Board

iv. Management information

v. Committees of the Board:
The Regulation requires the formation of an audit committee as the least, of committees of the board.

vi. The appointment of directors

vii. Whistle-blowing

The SEC has also developed 'guidelines for auditors of public companies and SEC Licensees' in 2017. The guidelines provide an avenue for auditors to independently assess governance and other risk associated with financial market operators.

The guidelines deal with:

i. Requirements for Auditors
ii. Approval of appointment to audit
The guidelines require the prior approval of the SEC for the appointment of an auditor
iii. Independence of an auditor
iv. Removal and resignation of auditors
v. Auditor's report to SEC on certain matters:
The guidelines require an auditor to report material matters to SEC as stated in Section 7:

> *"Where an auditor, in the performance of its functions becomes aware of a matter which may negatively affect investor or shareholder interests, the auditor shall within seven days after becoming aware of that matter, lodge with the Commission a written report on the matter and send a copy of the report to the auditee firm".*

viii. Suspension or cancellation of registration

The Bank of Ghana (BOG) has been championing good governance practices of financial institutions. BOG does background checks of all directors of financial institutions. Each director is required to submit a police report and obtain prior approval of BOG before appointment. Directors of financial institutions are personally liable for the liabilities of their financial institutions in the event of liquidation. The "***Bank of Ghana BSD Form F/ P001 (Personality Notes Form)"*** is required to be filled by each director and certified by an auditor. This form captures the assets and liabilities of the person as well as their location. The form requires information about other directorships of the person. Ghana has an independent anti-money laundering body called the Financial Intelligence Centre. The FIC carries out anti-money laundering test on directors as well as related party transactions. BOG has regulations on related party transactions and disclosures requiring these matters to be disclosed in the annual financial statements.

6.1.2 Public Sector Governance Changes

Ghana passed a new law for financial management for the public sector in 2016 titled "Public Financial Management Act, 2016 (Act 912). This law among others seeks to promote good governance in the management of public funds. Ghana has attained a lower-middle income status with dwindling donor support and grants and therefore requires prudent financial management of public funds. Act 912 replaces the Financial Administration Act, 2003 (Act 654) which did not fully address public sector governance issues on finance.

Act 912 provides regulations on the following governance issues:

i. Responsibilities and roles for public financial management
ii. Governance for public debt management
iii. Audit committee composition and size
iv. Formation and duty of governing boards of public corporations and state-owned enterprises
v. Removal of directors of a public corporation

The Internal Audit Agency of Ghana has developed guidelines for the operations of audit committees in the public sector to operationalize the requirements of Act 912. The guidelines deal with audit committee composition, size, training and orientation, remuneration and reporting requirements for public sector entities.

The Public Services Commission of Ghana has also developed a guideline for governance in the public sector. The guideline is called "Corporate Governance Manual for Governing Boards/Councils of the Ghana Public Services". According to Article 190 (1) of the 1992 Constitution of Ghana, the public sector comprises:

i. the Civil Service
ii. the Judicial Service
iii. the Audit Service
iv. the Education Service
v. the Prisons Service
vi. the Parliamentary Service
vii. the Health Service
viii. the Statistical Service
ix. the National Fire Service
x. the Customs, Excise and Preventive Service
xi. the Internal Revenue Service
xii. the Police Service
xiii. the Immigration Service
xiv. the Legal Service
xv. Public corporations other than those set up as commercial ventures
xvi. Public services established by this Constitution
xvii. Such other public services as Parliament may by law prescribe

The manual addresses the following governance issues:

i. Corporate Governance Architecture
ii. Appointment, Induction and Tenure:
The manual provides the topics to be covered in such training.

iii. Responsibilities and Duties
iv. Meetings
v. Committees:
Human Resource, Administration and Legal Committee, Finance Committee and Audit Committee must be formed at the least.
vi. Remuneration
vii. Corporate Reporting and Compliance
viii. Board/Council – CEO Relationship
ix. Evaluation of the Board/Council
x. Evaluation of the Chief Executive Officer
xi. Ethical Standards
xii. Risk Management
xiii. Social Accountability

6.2 Private Companies and SMEs Corporate Governance

Ghana has no guidelines on governance for private companies and small and medium enterprises. The governance practices of these organizations are voluntary and based on the preferences of the shareholders. Family owned private companies are noted for none or weak governance structures. The directors are usually spouses or friends with considerable level of informality with governance matters. These companies usually have no board and governance is in the hands of the chief executive who usually might be the majority or sole shareholder. This informality leads to one-man-show and results in poor governance and control structures. It can be estimated that close to 70 percent of family businesses have weak or no governance structures.

However, some private companies have adequate and effective boards. These companies have adopted the governance guidelines discussed above. These companies are not listed for a number of reasons but adhere to global best practices in governance. Board sizes of these companies range between three and eight members with separation of the role of the chairman and the chief executive officer.

6.3 Governance Education

Corporate governance courses were not topical in Ghana's education system at both the academic and professional levels prior to 2004. Today, corporate governance courses are taught at the postgraduate level and undergraduate courses have governance issues as topics. The Institute of Chartered Accountants has a paper at the professional level on governance titled 'Governance, Ethics and Strategy'. Several governance workshops continue to be held. Ghana has hosted several academic and professional conferences on governance and Ghanaians have been involved in governance conferences, workshops and meetings in and outside the continent of Africa. Academics in Ghana have published considerable refereed articles in reputable journals on governance. There are a number of books published in Ghana on corporate governance for tuition and practitioners by leading scholars. There are tertiary institutions with schools or departments offering programmes on governance for the award of postgraduate degrees or diplomas.

In the public sector, orientation and training are held for members of governing boards. These orientations and training are formalized and organized by the supervisory body of the units. Recently, the International Finance Corporation (IFC) has been organizing governance training for trainers in Ghana.

6.4 Corporate Governance Disclosures

Corporate governance issues are disclosed in the annual reports of public companies and financial institutions in Ghana. Annual reports disclose the following governance issues:

i. The pictures and names of members of the board
ii. The profile of board members
iii. Biological information (ages) of board members
iv. Chairman's statement
v. Report of the board to shareholders
vi. Corporate governance:
This section usually covers duties of the board, committees of the board, internal audit and controls and significant related party exposures.
vii. Financial risk management

6.5 Loopholes in Ghana's Corporate Governance Practices

Corporate governance in Ghana despite the positive practices above, has some bottlenecks, which must be addressed. The right of shareholders has not been adequately addressed. The capital market is stagnated because of the perception of poor dividends yet directors continue to receive huge fees and compensations. Despite the SEC Code 2010, little has been done to promote the rights of shareholders as required. Minority and individual shareholders have minimal rights and the major means of engaging directors is during the annual general meetings.

Secondly, gender issues are not addressed in Ghana's corporate governance regulations. Little has been done to promote gender issues in terms of representation and composition of boards. There is no national agenda to promote gender representation on boards of public sector entities. However, Ghana has an agenda to promote female representation in politics and government ministerial appointments and appointments as heads of public sector organisations. Currently, the Chief Justice is a female as well as a couple of ministers of the state.

Enforcement of regulation is another issue Ghana must deal with to promote corporate governance. The ROSC 2004 bemoaned the abysmal enforcement of standards and codes in Ghana. The situation has not significantly changed. The Companies Act gives at least a minimum standard of corporate governance. Nevertheless, majority of particularly private companies could be faulted for non-compliance of the governance rules in the Act. The Registrar of Companies can be held accountable for the lack of enforcement of these rules. Also, the right of shareholders to information is constantly abused because most companies fail to prepare annual reports and to file them with the Registrar General's Department as required by the Act.

The effectiveness of governing boards is a further issue Ghana must tackle to maximize the expected benefits and purpose of boards. Despite the requirement to measure the effectiveness of boards in the public sector, little is done to enforce this and to act when it is done. A similar requirement is not found in the SEC 2010 regulations, so assessing board effectiveness is left at the discretion of the company. Board effectiveness is found to contribute to performance. Ghana must develop guidelines to measure, promote and maximize the benefits of boards.

Compensation and earnings management are other matters lacking clarity and guidelines in Ghana's governance architecture. Currently, all the guidelines provide for the need to compensate board members. However, they fall short of providing the principles for determining the quantum of compensations and measures to check earnings management. Compensation in corporate governance is mostly not performance-based and shareholders are left at the mercy and discretion of the board. The capital market is not active due to low dividend payouts yet disclosures on directors' compensation continue to sour.

6.6 Conclusion and Implications

Ghana has seen considerable changes in corporate governance practices over the last decade in both the private and public sectors.

The SEC Code 2010 and the draft governance regulation 2017 have proven to be a good attempt to promote governance in the private sector. However, private companies do not have any such regulation. The statutory supervisory bodies such as the Bank of Ghana also promote governance in their licensing processes. The Public Financial Management Act and Corporate Governance Manual for Governing Boards/Councils of the Ghana Public Services have also provided adequate guidelines on governance for the public sector.

Corporate governance culture has been inculcated through education, training, research and publications. Ghana has seen a rise in books, workshops, conferences, postgraduate programmes and courses on governance. There are on-going collaborations on corporate governance in and outside the African continent.

In-spite of the above, governance issues suffer from poor protection of shareholders' rights, weak enforcement of governance regulations, failure to measure governance effectiveness and lack of clarity on board compensation. As a result, the capital market has a poor image and perception of boards acting for personal gains.

There is the need for regulators to establish desks or offices to champion or enforce governance rules in both the private and public sectors. The Ministry of Trade and Industry in collaboration with the private sector can consider instituting annual corporate governance awards to promote best practices.

REFERENCES

Ghana (1992). *1992 Constitution.* Accra: Ghana Publishing Corporation Ministry of Finance (2016). *Public Financial Management Act, 2016 (Act 912).* Accra: Ghana Publishing Corporation

Public Services Commission (2015). *Corporate Governance Manual for Governing Boards/Councils of the Ghana Public Services*. Accra: Public Services Commission

Securities and Exchange Commission (2010). *Corporate Governance Guidelines on Best Practices.* Accra: Securities and Exchange Commission

World Bank/IFAC (2004). *Report on Observance of Standards and Codes-Ghana Accounting and Auditing*. New York: World Bank/IFAC.

Chapter 7

Corporate Governance in Malawi

Andrew Munthopa Lipunga

7.0 Introduction

This chapter presents the evolution of corporate governance in Malawi. It highlights the strides made and the challenges that should be addressed in order for the country to have a robust system of corporate governance. The next section presents the historical background of the country, followed by a review of key developments in the evolution of the corporate governance framework since 1984. The remaining sections provide an analysis of the challenges at macro and micro levels, and the concluding remarks.

7.1 Historical Background

Malawi is a Sub-Saharan African country and one of the poorest on earth (Department of State, 2016:4). It was declared a British protectorate in 1891 (Nkhata, 2014:1), and in 1953 was made part of the federation of Southern Rhodesia (Zimbabwe) and Northern Rhodesia (Zambia) and was still under the British rule. The federation was broken in 1963 following nationalist movements that were led by, amongst other leaders, the founding father of the country, Dr Hastings Kamuzu Banda. The dissolution of the federation led to the independence of the country in 1964 and Dr

Banda became the first president. He ruled the country for 30 years under a one party dictatorship until 1994. In the prior year, Malawians through a referendum had voted for multiparty democracy. This led to general elections in 1994 that ushered another party into power. Since then, four presidents have ruled Malawi.

This background has a long lasting legacy that has important impact on the development of the system of corporate governance of the country in its current form and in the understanding of the same. Essentially the declaration of the protectorate in 1891 was an attempt to form a state of Malawi into a single centralised entity (Nkhata, 2014:1). However this was not a reflection of "the interests of the majority of the people of Malawi and was merely designed to further the interests of the colonial regime" (Nkhata, 2014:1), particularly to consolidate their control over the country's resources. Accordingly a few politically connected individuals – colonialists, controlled the country and its economic resources including public entities. Although the country got independence in 1964, unfortunately such was the legacy that was passed on to the new indigenous regime. Under the 30-year dictatorship rule of Dr Banda "although a formal division of power existed, the presidency was the only institution of import" accordingly "power was exceedingly concentrated in his hands" (Brown, 2008:187-227). As a result, it was the President and his closest associates, as well as most top politicians and civil servants that were in control of the country, including the public entities (Brown, 2008:187-227).

The change from dictatorship to democracy in 1994, did not help either. Even though the constitution "created an institutional configuration that balance the three branches of power—the executive, legislative and judiciary—and their constitutionally prescribed roles, nonetheless, the executive remains excessively strong" (Brown, 2008:187-227). This is attributed to the lack of consultation, participation, and representation that characterised the adoption of the new Constitution (Nkhata, 2014:8). The adoption was facilitated by a National Consultative Council that was made up of representatives of political parties with no popular mandate from the people (Nkhata, 2014:8). Accordingly the interests of the majority were not taken into account (Nkhata, 2014:8). Thus even now the President still retains much power and it is the President and close political allies that direct and control public entities.

On the other hand, due to the past history with Britain and the continued relationship as one of the major donors and through the commonwealth, the governance system of Malawi mirrors that of Britain. The country's legal system is based on English common law (World Bank, 2007:4). In effect, the corporate law and revision thereof follows developments of the British counterpart. Furthermore corporate governance guidelines were modelled on Commonwealth Association for Corporate Governance (CACG) corporate governance guidelines and other Anglo-American based corporate governance codes. Thus the corporate governance system of the country especially in the public sector is framed and continues to develop based on the legacy of the country's past history.

7.3 Development of corporate governance in Malawi

It is said that the phrase 'corporate governance' came into vogue in the 1970s in United States (Cheffins, 2013:1). However, in Malawi it got publicity in the 1990s, the same time that it became the subject of worldwide debate (Cheffins, 2013:2). Before that corporate governance practices existed through the various enacted laws and other corporate rules and regulations. Thus, development of the concept did not start with the formal recognition of the term or adoption of the code of corporate governance. **Table 1** provides a summary of significant developments undertaken to promote good standards of corporate governance in both public and private sectors over the years since 1984.

Table 1: Significant Corporate Governance milestones of the years

Enactment of Companies Act	1984
Enactment of Capital Market Development Act	1990
Enactment of Corrupt Practices Act	1995
Establishment of Malawi Stock Exchange	1996
Establishment of Anti-Corruption Bureau	1998
Launching of the Code of Best Practice for Corporate Governance in Malawi	2001
Enactment of Public Finance Management Act	2003
Enactment of Public Audit Act	2003
Establishment of Institute of Directors of Malawi	2003
Publication of ROSC Assessment – Corporate Governance	2007
Launching of Malawi Code II Code of Best Practice for Corporate Governance in Malawi: Overarching Provisions	2010
Launching of Corporate Governance guidelines for banks	2010
Enactment of Securities Act	2010
Launching of Sector specific Guidelines for: 1. Listed companies 2. Parastatal organisations and state owned entities 3. Incorporated Micro, Small and Medium Enterprises 4. Unincorporated Micro, Small and Medium Enterprises 5. Cooperatives, associations and membership-based organisations	2011
Launching of minimum standards on corporate governance for insurance companies	2011
Enactment of New Companies Act	2013
Establishment of Public Service Reform Commission (PSRC)	2014

Essentially, the enactment of the Companies Act in 1984 formally heralded good corporate governance in the country in the most recent past. The Companies Act was based on British company law, in particular the 1948 version (World Bank, 2007:4). It is this Act together with a string of legislations that established various public entities that built the system of corporate governance in private and public entities before the adoption of a formal code.

The enactment of the capital market development Act in 1990 was another key development worth recognition. The Act provided the legal framework for the development of the capital market. It facilitated the establishment of the Malawi Stock Exchange (MSE) in 1996. The establishment of a stock market was in itself another important development in the evolution of corporate governance due to the additional rules and regulations that it imposed on the listed companies relating to governance practices (Haque, Arun, & Kirkpatrick, 2008). This was manifested in the country in that, even before the country adopted its own code, MSE listing rules required listed companies to disclose in their annual reports the extent of their compliance or non-compliance with the Cadbury or King Report on Corporate Governance (MSE, 2008:11). Thus the listing rules helped to bring corporate governance to the fore and align the country with international best practice.

Another important development was the enactment of the Corrupt Practices Act in the same year, 1995. This led to the establishment of the Malawi Anti-Corruption Bureau in 1998 with the vision of eliminating all forms of corruption (Nampota, 2008:1). This was significant as corruption is one of the problems that hinder good governance in most developing countries like Malawi. It is recognised that "an active approach to prevention, setting the tone from the top, is the way to ensure good corporate governance" (Applied Corporate Governance, 2016, July 13). The Corrupt Practices Act gives the Bureau power to investigate corrupt cases, educate and enlist public support against corrupt practices and prevent corruption (Nampota, 2008:2-3). Thus the Act attempts to establish a strong ethical basis in the governance system of the country, which is important because corruption erodes the quality of public governance and that negatively affects corporate governance (Chakraborty, 2015:85).

A remarkable milestone was reached in 2001 when the first national code of corporate governance was adopted. The process had originated in 1997 at a three-day conference organised by the Society of Accountants in Malawi (SOCAM), where the first ever open discussion on corporate governance in the country took place (SOCAM, 2001:1). Based on the resolutions at the conference, SOCAM facilitated the formulation of a taskforce in 1998 that fulfilled its mandate in 2001. The taskforce comprised captains of industry representing all relevant professional and industrial associations and representatives from Malawi Stock Exchange, Anti-Corruption Bureau and relevant government agencies.

The code was developed based on South Africa's King Report, the Commonwealth Association for Corporate Governance (CACG) Guidelines and the Kenyan Principles of Corporate Governance (SOCAM, 2001:2). The coverage of the code included the following topics: compliance, board of directors, chairman, non-executive directors, appointment of directors, directors' remuneration, training and development of directors, company secretary, stakeholder communications, financial reporting and auditing, worker participation, and code of ethics (SOCAM, 2001:1-8). The code, which was developed on 'comply or explain' basis was expected to be applied by all organisations both in private and public sectors irrespective of the size.

As a way of facilitating successful implementation of the code, the task force also recommended: (1) review of the then Companies Act of 1984 to reflect the changing environment; (2) establishment of an institute of directors; (3) capacity building at the registrar general's office on dealing with non-compliance with the Companies Act; and (4) active engagement of all professional bodies and other regulatory bodies in the promotion of good corporate governance environment (SOCAM, 2001:8). SOCAM became the custodian of the code until 2003 when the Institute of Directors of Malawi was established.

Following the recommendations of the Task Force, the Institute of Directors of Malawi (IOD) was established in 2003 facilitated by SOCAM. Its mission was to promote good corporate governance in Malawi (Lipunga, 2014:247). Its objectives include being the custodian and promoter of the code, educating directors and senior managers on issues of governance, monitoring conduciveness of legislations, speaking for directors and executives in public affairs; and upholding good business practices (Lipunga, 2014:247).

In the same year, 2003, the Public Finance Management Act and Public Audit Act were enacted. These were notable developments to corporate governance in the public sector because, the Public Finance Management Act of 2003: (1) attempted to clarify the responsibilities of the Minister, Secretary of Treasury and Controlling Officers; (2) stipulated the development and maintenance of an effective system of internal controls including an effective internal audit function in public entities; (3) required preparation of annual reports by public entities within four months after year end that adhere to generally accepted accounting practice; (4) mandated the auditing of financial reports in public entities; and (5) required statutory bodies to prepare and furnish the minister of finance and the line ministers with performance and management report together with financial statements within four months of the end of the financial year of the body (Government of Malawi, 2003B:1-58). On the other hand, the Public Audit Act of 2003, legislated the establishment of the office of Auditor General with the duty and power to audit every ministry, statutory offices, agencies, boards, commissions and bureau of the Government, and public funds received by non-profit organizations, including relevant international organizations. Furthermore the Act established the Public Accounts Committee of Parliament with the primary aim of promoting transparency and accountability of government. The Act mandated the public accounts committee to (1) make information available to enable Parliament make informed decisions on the management of public resources; (2) promote accountability of Ministers; (3) hold ministries, agencies and other entities controlled by government accountable for the economy, efficiency and effectiveness with which they use public resources; and (4) promote the accountability of statutory bodies and those other persons charged with the management of public resources (Government of Malawi, 2003A:1-21).

Thus the two Acts legislated the application and enforcement of the key corporate governance principles of transparency, accountability and responsibility in the public sector. Furthermore the legislations set up the oversight structure for public sector entities. As exhibited in **Figure 1** oversight for public entities is supposed to be done by the line ministries, ministry of finance, department of statutory corporations (in case of statutory organisations) and the parliament through the public accounts committee.

Figure 1: Oversight structure of public entities in Malawi

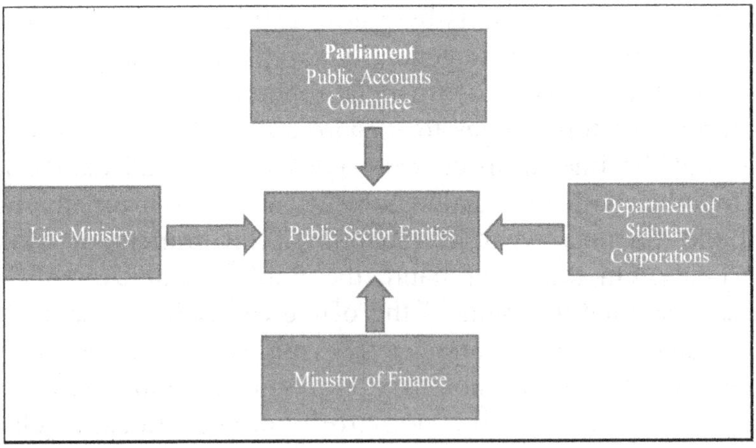

The publication of the Report on Observance of Standards and Code (ROSC) on Corporate Governance in 2007 was the other important development that helped the country strengthen its governance system. The report highlighted significant gaps and recommended improvement measures. The short-term measures recommended included: enhancement of support to the Institute of Directors, revision of the code of corporate governance, revision of the statutory framework for accounting and auditing, development of a strategy for improving Parastatal governance, and development of enforcement strategy for good corporate governance practices (World Bank, 2007:10). The long-term measures included revision of the Companies Act, non-financial disclosure framework for listed companies, securities law and listing requirements (World Bank, 2007:10).

Essentially, the report found that the Institute of Directors had severe difficulties to become financially sustainable, hence was incapable to operate effectively. As a result, it needed support from both public and private sectors. Furthermore, it found that, the code had no local support as such companies kept making their corporate governance compliance statements with reference to external codes. Even MSE's listing rules required corporate governance disclosure by listed companies to be made against the Cadbury or King Report (MSE, 2008:11; IOD, 2010:4). Besides, the Companies Act of 1984 was found not in harmony with best practices and responsiveness to dynamics of the environment. Accordingly the report particularly recommended updating of responsibilities of the board in line with international best practice, harmonising of accounting requirements with modern accounting standards; updating the contents of the directors' report; modernising the regime for share recordkeeping and definitions of direct and indirect ownership; updating creditor rights and the role of the board during insolvency; and harmonizing takeover regulation in the Companies Act with the Listing Regulations (World Bank, 2007:6-7).

Following the recommendations of ROSC, the IOD constituted a National Corporate Governance Review Committee (NCGRC) to review the code. The committee comprised members from all relevant professional and industrial associations and it received technical assistance from the International Finance Corporation –

Global Corporate Governance Forum (IOD, 2010:8). This led to the publication of the second code in 2010 called "Malawi Code II Code of best practice for corporate governance in Malawi: Overarching provisions" which is currently in force. The publication of the code was the initial phase of the development of a comprehensive set of corporate governance rules. The other phase would involve development of sector specific guidelines intended to provide the necessary interpretation, best practice and, more stringent requirements for implementation in a particular sector (IOD, 2010:5). Thus the code contained only the general principles applicable to all types of organisations (IOD, 2010:5). The new code further incorporated some key issues that were not in the old code such as risk management, board evaluation, good corporate citizenship, related party transactions and sustainability (IOD, 2010:4).

Another development in the same year 2010 was the issuance of corporate governance guidelines for banks by the Central Bank of Malawi. The guidelines were fully recognised by the IOD as sector specific guidelines for the banking sector (IOD, 2010:5). The guidelines were issued in lieu of a directive (RBM, 2010:2), hence they are statutory in nature however on "comply or explain" basis (RBM, 2010:21). The guidelines that are still in force are comprehensive and comparable to those of the neighbouring countries. Their publication gave impetus to the promotion of corporate governance as they are applicable to an influential sector in the economy and they are rigorously enforced by the Central Bank.

Further development in the year was of the enactment of a long awaited Securities Act, which superseded the Capital Markets Development Act. ROSC in 2007 found the Capital Markets Development Act of 1990 to be out-dated and recommended its complete reformation (World Bank, 2007:17). In line with the recommendation of ROSC, the new Act enhanced the enforcement powers of the Central Bank over the stock exchange and its members and also of the stock exchange to ensure compliance of its rules and to take measures against non-compliance (Government of Malawi, 2010:1-42).

In 2011, the IOD issued five sector specific corporate governance guidelines for listed companies: Parastatal organisations and state owned entities, incorporated Micro, Small and Medium Enterprises; unincorporated Micro, Small and Medium Enterprises, cooperatives, associations and membership-based organisations. Furthermore, the Central Bank issued the minimum standards on corporate governance for insurance companies. These minimum standards were issued as a directive (RBM, 2011:44). The development of these six sector specific guidelines completed the development of corporate governance code as envisaged by the NCGRC.

Another remarkable development took place in 2013, which was the enactment of a new Companies Act. Amongst the important new provisions, the Act recognised the importance of corporate governance. Sections 70 and 184 recognised compliance with the code of corporate governance as a core duty of the directors (Government of Malawi, 2013:52,105). However according to Section 184(3) the stipulations are 'directory in nature'. Nevertheless the Act empowers the Courts, Registrar of Companies or any authority to consider the codes in interpreting and applying any of the provisions of the Companies Act. The new Act has raised the profile of corporate governance in the country. This is expected to increase the rate of compliance with the codes and give impetus to the IOD to promote widespread adoption of codes.

Finally, in 2014 the Government took a bold step towards good governance in the public sector by setting up the Public Service Reforms Commission (PSRC). This was launched in fulfilment of a campaign promise made by the President. The launch was remarkable due to, among other things, its scope and the calibre of commissioners that attended. The Vice President of the country, an accomplished corporate executive, chaired the event. All the commissioners were reputable private and public sector gurus and activists. Among them was the former Chief Executive Officer of the Institute of Chartered Accountants in Malawi (ICAM) whose predecessor, SOCAM, has been on the forefront championing corporate governance in the country. Furthermore she was part of the team that developed the current codes of corporate governance. The commission has since submitted its final report, which recommends sweeping reforms amongst which is the de-politicisation and entrenchment of ethics, integrity and discipline of the public service (Government of Malawi, 2015:39-40).

The discussion above, indicates that corporate governance in Malawi has evolved from simply being based on the provisions of legislations (up until 2001), to its current form in which the code is well recognised by corporate law. Furthermore, it is being supported by a number of institutions. The discussion further indicates that key institutions that have been actively engaged in the development and promotion of corporate governance include the government particularly through enacting relevant legislations and its oversight role in some of its agencies in public and private sector entities. An important government agency worth mentioning is the Central Bank. Apart from its oversight role over the financial sector, it established the Stock Exchange and has issued comprehensive corporate governance guidelines for banks and insurance firms.

The Institute of Chartered Accountants of Malawi (ICAM) is another statutory body whose predecessor, SOCAM, facilitated the first open discussion on corporate governance that led to the development of the first national code and establishment of the IOD. Currently, ICAM has also incorporated corporate governance in its educational programmes at all levels (i.e. certificate, diploma and professional levels). Finally, the IOD has been instrumental in developing the second code of corporate governance and the sector specific guidelines while the Malawi Stock Exchange has been promoting corporate governance among listed companies.

7.4 Issues and challenges at macro level

As discussed above, Malawi has made strides in promoting corporate governance. The country has a legal framework that provides for and supports good corporate governance. The latest being the enactment of the new Companies Act in 2013, that explicitly requires directors of private and public companies to comply with corporate governance codes in force in Malawi. Although it is on directory basis, nevertheless this inclusion has helped to raise the profile of the concept. The burden now rests on the IOD to seize the opportunity to have the concept entrenched in the conduct of all organisations. Besides, the Public Finance Management Act provides also for greater accountability and transparency and oversight over public sector entities.

The major challenge however is sheer lack of resource support for the institutions charged with promoting good governance. For instance, IOD had been operating with only two key staff and currently it is not operational after the death of its executive director and resignation of the other officer. When it was operational, it was unable to accomplish most of its mandate due to poor capacity. It was struggling financially, to the extent that it remained closely linked (indistinguishable) to SOCAM (World Bank, 2007:39). ROSC (2007) reported the fact and appealed for greater support, which seems not to have materialised up to date. Due to the current state of the IOD, practically the country does not have a body with the mandate to promote and monitor compliance with the codes of corporate governance.

Besides, the parliament's public accounts committee suffers also from lack of funding (SADCOPAC, 2009:13; Gwede, 2016, April 9). Furthermore, the executive arm of the government does not respect the committee's recommendations (SADCOPAC, 2009:13). Above and beyond, the effectiveness of the committee appears to be contingent on its composition. If the majority of the members are from the opposition, the committee appears 'busy and objective' otherwise if they are from the ruling party it is 'passive and less critical' perhaps for fear of denting the image of their party.

As exhibited in **Figure 1**, the line ministries, ministry of finance and department of statutory corporations also oversee public sector entities. Furthermore there is a department of public sector reforms management that is responsible for initiating and managing reforms in the public sector. To date, the effectiveness of these bodies is questionable as little can be shown as their impact. Their work seems to be limited to merely producing good policy documents. The oversight function of these bodies seem to be constrained by politics involved in the running of public sector entities. This is so because the President appoints majority of the board of directors in public entities who are influential politicians or politically connected persons or senior civil servants (Department of State, 2016:23). As a result, making them accountable is quite challenging for they are more powerful than the professional civil servants.

Politically, the willingness to promote good corporate governance practices has generally been on paper and in words. Reforming the public sector has been a song since the rebirth of the country in 1964. So far there have been 79 failed reformation attempts (State House Malawi, 2015, February 12; Chingaipe, 2015, January 15), attributable to lack of political will (Government of Malawi, 2015:50). Although political leadership may have approved the reforms they appear to have other motives high on the list such as to be seen to be doing something. However the launch of the Public Service Reforms Commission in 2014 was seen as a serious demonstration by the political leadership of the willingness to embrace change. It gave a sigh of hope for good governance in the public sector. As already stated, the commission has since presented its final report and was disbanded at the expiry of its mandate. It was hoped that implementation of the recommendations will be overseen by the same commission or at least the Office of the Vice President. However, the responsibilities have been withdrawn and reverted to the Office of the President and Cabinet, hence there is significant level of scepticism about whether the reforms will actually be implemented (Chingaipe, 2015, January 15).

Other general challenges include the high levels of political corruption (Nawaz, 2012:1). This is critical because the quality of public governance tends to be eroded by corruption that in turn negatively affects corporate governance (Chakraborty, 2015:85). Ironically Malawi is seen to have strong anti-corruption laws and institutions and initiatives by the private sector that complement the governmental efforts (Nawaz, 2012:1). However corruption continues to be a problem in the public sector (Department of State, 2016:24), due to "a significant gap between law and practice" (Nawaz, 2012:1). The anti-corruption fight suffers greatly as the lead institution, the Anti-Corruption Bureau, lacks adequate funding, human resources and sufficient independence (Nawaz, 2012:1-2; Strasser, 2016:303-337).

Another general challenge is the lack of a sense of urgency for change in regulatory instruments with the passage of time. Connected to this challenge is lack of internal drive and decisive pro-activeness to reforms. Generally development of legislations, codes and other regulatory instruments and their subsequent revision take time to be initiated and completed and they are normally externally triggered. There seems to be no internal mechanisms that monitor their conduciveness and environmental changes and subsequently initiate necessary reforms and revisions. A good example is the first code of corporate governance itself. It was revised only after the ROSC, in spite of clear indications that organisations, including the stock market had no confidence in it.

The other general challenge is the tendency of modelling the governance framework on other countries' frameworks in spite of intrinsic differences. Basically an effective corporate governance system is a product of legal, regulatory, and institutional environments as well as hierarchal, cultural, historical and political factors (Donker & Zahir, 2008:83-93; Maher & Andersson , 2009:4,5). Failure to take cognisance of environmental factors results in misalignment and eventual non-compliance. The current codes are stakeholder-oriented based on African 'inclusive' values. However the underlying corporate legal framework is shareholder-oriented as it is based on English law (Khomba, 2011:228) causing some internal disharmony in the system.

The provisions regarding the production of annual reports and accounts are classic examples of the disharmony by the Companies Act. It is only mandatory for public and state owned companies. Small companies are exempted, while the other private and state owned companies get exempted if their 'shareholders' decide otherwise and by a published notice respectively (Government of Malawi, 2013:1-208). Exempted entities are only required to produce financial statements. The exemption effectively exempts them from making corporate governance disclosures. The exemption provision may make sense in 'exclusive' settings that emphasise on shareholder primacy. However they pose a problem to the "inclusive" African culture that commits explicitly to serve the interests of both shareholders and non-shareholding stakeholders (Khomba, 2011:208).

7.5 Issues and challenges at micro level

The governance framework explicitly requires all organisations to state in their annual or directors' reports whether the Code has been adhered to or, if not, explain with reasons in what respects it has not been adhered to (IOD, 2010:11). However the evaluation of the implementation of the requirement by organisations is a challenge due to lack of enforcement and monitoring mechanisms. The code is intrinsically voluntary. Accordingly, although compliance is a duty for directors as provided in the Companies Act, strict compliance cannot be enforced since it is on directory basis. The situation is exacerbated by the lack of institutional capacity by the IOD and office of Registrar's General to take measures with organisations that fail to comply with the Companies Act (World

Bank, 2007:32). The case is different for the 13 listed companies, 10 banks and insurance firms, because over and above the provisions of the code, their regulators require production of annual reports with explicit requirement for corporate governance disclosures therein. The greater majority of the other organisations in the private sector are likely to report based on the minimum requirements of the Companies Act, which would simply be the production of financial statements without making corporate governance disclosures.

Major challenges are also in the public sector at the organisational level. Although the Public Finance Management Act provides for good governance practices that include preparation of annual reports and auditing (Government of Malawi, 2003B:1-58), in practice not all public entities produce the annual reports on time nor are they always audited by independent professional audit firms (Department of State, 2016:22). Another critical governance challenge is the appointment and composition of the boards of directors of public entities. The national code requires members of the board to be of sufficient calibre to bring independent judgement (IOD, 2010:14). However, the boards are composed of politicians and senior government officers appointed by the President (Department of State, 2016:23). As narrated above, this makes government agencies charged with oversight over the entities ineffective. The ministers who would have been in a better position to rescue the situation have existing political connections with the board of directors, and this affects their role. This leaves the parliament's public accounts committee with the greater oversight burden. Unfortunately, as evident from the foregoing, the committee's effectiveness seems to depend on the make-up of its membership. If majority of its membership are from the ruling party, oversight gets affected as they monitor their own clones. On the other hand, if majority of its membership are from the opposition party, rigorous assessment may be interpreted as political manoeuvring.

Furthermore, the other governance weakness in the composition of the boards of public entities is the membership of senior government officials representing government departments as ex-officio/non-voting members (Department of State, 2016:23). These are usually the controlling officers of the department of statutory corporations, line ministries and ministry of finance. Hence they are the same officials charged with the oversight over the same entities according to Public Finance management Act. Thus their participation, as well as those of politicians, creates a perceived and/or real conflict of interest (Department of State, 2016:23) that needs to be addressed.

Another significant challenge is the way the top management in ministries and government departments are appointed. The final report of the PSRC noted that the Public Service Act of 1994 vests

> "the power to appoint senior officers… of positions above Under Secretary … in the President and that there has been no provision for a due process of competitive interviews. As a result, selection, appointments and promotions have mostly been based or seen as based on connections to one's superiors or politicians rather than merit" (Government of Malawi, 2015:39).

Thus the selection process even of top civil servants is marred by elements of "cronyism" and "tokenism" that corporate governance best practice seeks to avoid (CACG, 1999:8). These senior officers are important to corporate governance in the entire public sector, as they are the ones charged with oversight of other public entities under their respective ministries as well as serving on their boards. Accordingly the PSRC recommended "the President's powers of appointing should be limited to the approval of the results of the selection process from Public Service Commission" (Government of Malawi, 2015:40).

The analysis above suggests that in the current set up, implementation of good corporate governance practices in public entities in Malawi hinges on political leadership, in particular the President. Thus it is recognised that implementation is not an issue of technicality in administrative reforms but a profoundly political process (Ng'oma, 2010:70). The challenge is that process would require strenuous and protracted negotiations and bargaining as such reforms have the effect of redistributing power (Ng'oma, 2010:70).

7.6 Conclusion

The Chapter suggests that Malawi has moved from a corporate governance system based on the legal framework to having a national code recognised as well. However, despite having a legal system that supports good corporate governance and the national code, enforcement, monitoring and prompt revision of the instruments seem to be the major challenge. In particular the country is failing to support responsible institutions with adequate resources, which makes it is difficult to assess the overall state of implementation of good corporate governance practices at micro level except for listed companies, banks and insurance firms.

In the public sector, the political order seems to be the major factor in the enforcement of good corporate governance practices. Power is concentrated in the President, as such, implementation depends on adoption of the same by the political leadership. This is the reason the PSRC recommended that "the President should be the sole sponsor of reforms at national level" (Government of Malawi, 2015:50). In harmony, Chingaipe (2015, January 15) asserted "the missing link so far has been the recognition that Public Service Reforms is not just a technocratic process, it is a profoundly political one too and needs both optimal technocratic and political leadership".

REFERENCES

Applied Corporate Governance. (2016, July 13). Retrieved March 23, 2017, from http://blog.applied-corporate-governance.com/ethics/corruption-and-corporate-governance/

Brown, S. (2008). The Transition from a Personal Dictatorship: Democratization and the Legacy of the Past in Malawi. In S. W. Nasong'o (Ed.), *The African Search for Stable Forms of Statehood: Essays in Political Criticism* (pp. 187-227). New York: Edwin Mellen Press. Retrieved March 20, 2017, from http://stephenbrown.xyz/wp-content/uploads/Stephen-Brown-Malawi.pdf.

Chakraborty, I. (2015). How Does Corruption Influence Corporate Governance? A Cross-Country Perspective. *ICRA BULLETIN Money & Finance*, 67-88.

Cheffins, B. R. (2013). The History of Corporate Governance. In M. Wright, D. S. Siegel, K. Keasey, & I. Filatotchev (Eds.), *The Oxford Handbook of Corporate Governance.* doi:10.1093/oxfordhb/9780199642007.013.0003

Chingaipe, H. (2015). Researching the politics of public sector reforms in Malawi. Retrieved March 27, 2017, from http://www.effective-states.org/researching-the-politics-of-public-sector-reforms-in-malawi/

Commonwealth Association for Corporate Governance (CACG). (1999). *CACG Guidelines: Principles for Corporate Governance in Commonwealth.* Retrieved from http://www.ecseonline.com/PDF/CACG%20Guidelines%20-%20Principles%20for%20Corporate%20Governance%20in%20the%20Commonwealth.pdf

Department of State. (2016). *Doing Business in Malawi: 2016 Country Commercial Guide for U.S. Companies.* Retrieved from http://photos.state.gov/libraries/malawi/217630/PDFs/Malawi%20CCG%202016.pdf

Donker, H., & Zahir , S. (2008). Towards an impartial and effective corporate governance rating system. *Corporate Governance, 8*(1), 83-93.

Goverment of Malawi. (2015). *Public Service Reforms Commission Final Report.* Retrieved March 13, 2015, from https://info.undp.org/docs/pdc/Documents/MWI/Malawi%20Public%20Service%20Reform%20Report(1)%20(2).pdf

Government of Malawi. (2003A). Public Audit Act.

Government of Malawi. (2003B). Public Finance Management Act.

Government of Malawi. (2010). *Securities Act.*

Government of Malawi. (2011). *Malawi Growth and Development Strategy II 2011-2016.* Retrieved from www.mw.one.un.org/wp-content/uploads/2014/04/Malawi-Growth-and-Dedvelopment-Strategy-MGDS-II.pdf

Government of Malawi. (2013). Companies Act, 2013. Government Printers Press.

Gwede, W. (2016, April 9). Malawi Parliament to prioritise funding of committees with oversight functions. *Nyasatimes*. Nyasatimes . Retrieved March 27, 2017, from http://www.nyasatimes.com/malawi-parliament-to-prioritise-funding-of-committees-with-oversight-functions/

Haque, F., Arun, T., & Kirkpatrick, C. (2008). Corporate governance and capital markets: a conceptual framework. *5(2)*. Retrieved March 31, 2017, from https://www.google.com/url?sa=t&rct=j&q=&esrc=s&source=web&cd=2&cad=rja&uact=8&ved=0ahUKEwjv4Kqhp4DTAhVHCMAKHXaYC0sQFggiMAE&url=https%3A%2F%2Fpureapps2.hw.ac.uk%2Fportal%2Ffiles%2F1920738%2FHaq

ue2012.pdf&usg=AFQjCNH85-c8FeHRJO29cdzbZRgz4bBXYg&bvm=bv.15132

Institute of Directors of Malawi (IOD). (2010). *The Malawi Code II Code of Best Practice for Corporate Governance in Malawi: Overarching Provisions.* Retrieved from http://www.ecgi.org/codes/documents/malawi_codeii_1jun2010_en.pdf

Khomba, J. K. (2011). *Redesigning the Balanced Scorecard Model: An African Perspective.* PhD Thesis, University of Pretoria, Faculty of Economic and Management Sciences, Pretoria. Retrieved March 2, 2017, from repository.up.ac.za/bitstream/handle/2263/28706/Complete.pdf?sequence=13

Lipunga, A. M. (2014, April). A Review of Disclosure Provisions in Malawian Codes of Corporate Governance. *International Journal of Business and Social Science, 5*(5), 246-256.

Maher, M., & Andersson , T. (2009). *Corporate governance: Effects on firm performance and economic growth.* Organisation for Economic Co-operation and Development. Retrieved March 16, 2017, from https://www.oecd.org/sti/ind/2090569.pdf.

Malawi Stock Exchange (MSE). (2008). *Malawi Stock Exchange Listing Requirements.* Retrieved from http://www.mse.co.mw/rules/MALAWI%20STOCK%20EXCHANGE%20LISTING%20REQUIREMENTS.pdf

Nampota, A. E. (2008). The Malawi Anti-Corruption Bureau. Retrieved March 7, 2017, from https://ancorage-net.transparencia.pt/wp-content/themes/twentyfourteen/assets/documentos/nampota.pdf

Nawaz, F. (2012). *Overview of corruption and anti-corruption in Malawi.* Transparency International. Retrieved March 27, 2017, from http://www.u4.no/publications/overview-of-corruption-and-anti-corruption-in-malawi/downloadasset/2829

Ng'oma, H. C. (2010). *Business and the State in Malawi: The politics of institutional formation, maintenance and change.* PhD Thesis, University of York, Department of Politics. Retrieved March 29, 2017, from https://www.google.com/url?sa=t&rct=j&q=&esrc=s&source=web&cd=1&cad=rja&uact=8&ved=0ahUKEwiEn57v_rSAhWMB8AKHcnwDi0QFggZMAA&url=http%3A%2F%2Fetheses.whiterose.ac.uk%2F1507%2F2%2FBUSINESS_AND_THE_STATE_IN_MALAWI_final.pdf&usg=AFQjCNFS0PoaRV2yvyNU9hDOkoaPty

Nkhata, M. J. (2014). *The Republic of Malawi Country Report.* Retrieved March 23, 2017, from http://www.icla.up.ac.za/images/country_reports/malawi_country_report.pdf

Reserve Bank of Malawi (RBM). (2010). Corporate Governance Guidelines. Retrieved from http://www.rbm.mw/documents/basu/Corporate%20Governance%20Guidelines.pdf

Reserve Bank of Malawi (RBM). (2011). Minimum standards on corporate governance for insurance companies. Retrieved from https://www.rbm.mw/Supervision/GetContentFile/?ContentID=3606

Society of Accountants in Malawi (SOCAM). (2001). Code of Best Practice for Corporate Governance in Malawi. Retrieved October 9, 2013, from http://www.track.unodc.org/LegalLibrary/LegalResources/Malawi/Laws/Code%20of%20Best%20Practice%20for%20Corporate%20Governance%20(2001).pdf

Southern Africa Development Community Organization of Public Accounts Committees (SADCOPAC). (2009). Malawi Public Accounts Committee faces inadequate funding. pp. 12-13.

Retrieved March 27, 2017, from http://www.sadcopac.org/modules/newsletters/index.php?action=downloadfile&filename=SADCOPAC%20Newsletter.pdf&directory=&PHPSESSID=ad4d8ae688168bb05fe96de45a02ee79

State House Malawi. (2015, February 12). *State Residences Malawi*. Retrieved March 16, 2017, from State Residences Malawi: http://www.statehouse.mw/apm-launches-public-service-reform/

Strasser, P. G. (2016). An Anti-Corruption Bureau's Inexorable Endeavor: A Study of Malawi's Cashgate Scandal. *Washington and Lee Law Review Online, 73*(1), 303-337. Retrieved March 7, 2017, from http://scholarlycommons.law.wlu.edu/cgi/viewcontent.cgi?article=1054&context=wlulr-online

World Bank. (2007). *Corporate Governance Country Assessment : Malawi. Report on the Observance of Standards and Codes (ROSC)*. World Bank, Washington DC. Retrieved March 9, 2017, from https://openknowledge.worldbank.org/handle/10986/20452 License: CC BY 3.0 IGO

Chapter 8

Corporate Governance in Mauritius

Jyoti Devi (Brinda) Mahadeo

8.0 Introduction

Governance extends well beyond the corporations. Governance is a country's attempt to distribute capital and resources in a fair and transparent fashion. At various points in time, different countries have utilised an assortment of institutional arrangements to control the people in whom the responsibility of resource and capital allocation has been placed. Mauritius applies a version of the free market economy as its economic model, whereby the State establishes the rules and makes available the physical and social infrastructure, whereas the private sector is responsible for wealth creation and employment. Mauritius has retained its position as the best country in Sub-Saharan Africa for doing business (Doing Business, 2017) and it is continuously leading the ranking on the Ibrahim Index of African Governance (IIAG) since the year 2006 (Mo Ibrahim Foundation, 2016). These excellent scores were sustained amid local headwinds and global shocks. The island of Mauritius has had its copious share of corporate scandals due to improper control prevalent in leading local companies such as Air Mauritius, Mauritius Commercial Bank/National Pensions Fund and, more recently, the British American Insurance Mauritius, which was considered by the current government as operating as a Ponzi Scheme.

By reviewing the developments in Corporate Governance in Mauritius, this chapter will endeavour to answer the following questions: i) how corporate governance developments have enhanced the strategic direction of national business enterprises, ii) determine how the country has attained international recognition as a priority destination for FDI and iii) establish whether the local enterprises are acting as responsible corporate citizens. Following this introduction, the first section will provide a background of the country beginning from its colonial origins, moving through the post-independence era to the current socio-economic times. The second section presents the moves introduced to promote standards of good corporate governance in the public and private arenas with particular emphasis on the development of the local Codes of corporate governance. Before concluding the chapter, the last section discusses the country's main achievements to attain global standards and successful economic and social development.

8.1 The Country in brief

Mauritius is a small island and developing country located in the Southern Indian Ocean. In sequence, the country has experienced colonialism under the French (1767 – 1810) followed by the British as a colony (1810 – 1968). Consequently, Mauritius has inherited a legal system that is a cross of the French–based law (contract and civil) and the British and Commonwealth legislation (common/company law). The island's colonial heritage has influenced property and business ownership (Lange, 2003). For instance, large family enterprises, which gained a great deal from the sugar business, were able to invest in other industries. Simultaneously, the distribution of land to sugar workers allowed the latter to aspire to a better life and gain access to public education. Hence, an active middle class emerged who were keen on white-collar jobs (Meisenhelder, 1997). The local routine and lifestyle of the population (descendants from India, China, Africa and Europe) remained closely linked to the French and British customs (Brautigam & Diolle, 2009).

Ostensibly, Mauritius is related to the African continent (through trade groupings and political memberships); however, many observers deny its similarity with African societies on the grounds of its island status and the immigrant origin of its people (Miles, 1999). In a comparison of colonial states performed Lange (2003) found more similarity between Mauritius and British Asia (Singapore, Hong Kong, and Malaysia) than British Africa. This similarity extends to being export-driven, relying on a small and highly skilled labour force; in addition, the country is administered by a broad national consensus indicative of the close association of political, administrative and economic elites (Brautigam & Diolle, 2009). Another significant differentiating factor is that colonialism was not brought from outside; instead, colonialism was built into the society of Mauritius (Houbert, 1981).

The grim predictions made by the Nobel Prize winner Professor Meade in 1960 never materialised (Meade, 1961), as the country bloomed into a successful nation within four decades and was dubbed the 'Mauritian Miracle' (Alladin, 1993). The country enjoys one of the highest standards of living in Africa due to the creative and apt moves engineered by the State. The most recent "Doing Business 2017" report, which was jointly prepared by the World Bank and the International Monetary Fund (IMF), ranked Mauritius as the best country of Africa (and 49th worldwide) for undertaking business. In recent decades, the Government has worked in accordance with the prescriptions of supra-national bodies (World Bank and IMF) to revamp the Corporate Regulatory Framework. The objective is to attract local and international investment in business projects. In 1989, the Stock Exchange of Mauritius (SEM) was created. The year 2001 was critical for the local corporate scene, as the Government doggedly introduced a series of interventions to propel the country forwards in its mission to attract investment. These interventions are discussed in detail in the following section.

The island nation has good international relations and is a member of the African, Caribbean and Pacific Group of States, the African Union, the British Commonwealth, the Common Market for Eastern and Southern Africa (COMESA), the Indian Ocean Rim Association, the New Partnership for Africa's Development (NEPAD), the Non-Aligned Movement, Organisation Internationale de la Francophonie, Southern African Development Community (SADC), the United Nations (UN) and the World Trade Organisation (WTO).

8.2 Corporate Governance Developments and Reforms

The steps taken by the Government from 2001 were foundation stones to modernise the country's corporate environment. The measures implemented were as follows: i) the passing of a new Companies Act in 2001, which replaced most of its predecessor (the Company Act, 1984). The new Act geared the law towards international business practices, such as defining directors' duties and delineating the procedures for calling meetings amongst others. ii) the introduction of International Financial Reporting Standards (IFRS). iii) the addition of new listing rules to the SEM. iv) the establishment of the National Committee on Corporate Governance (NCCG), and v) a request from the Government to the World Bank to undertake a Report on the Observance of Standards and Codes (ROSC) in Mauritius. The ROSC is a Corporate Governance Assessment, which forms part of the World Bank-IMF programme, which benchmarks the country's state of Corporate Governance against the OECD Principles of Corporate Governance.

8.2.1 ROSC 2002

The first ROSC report was prepared by Fremond and Gorlick; the recommendations were published in 2002. The report noted that the private sector was composed of a concentration of indigenous elite (one or two businesses) who controlled most of the economic sector or was ruled by state institutions with weak regulatory agencies, which were leading to economic inefficiencies. The World Bank, through the 2002 ROSC report, made five policy recommendations: i) obtaining the means to unlock shareholder value by strengthening valuations on the stock market, in part by implementing company legal reforms; ii) enhancing the capacity of regulators; (iii) implementing an Institute of Directors (IoD) and developing a corporate governance code; (iv) enhancing the regulation of the auditing profession; and (v) further attending to Global Business Companies (GBC)s. The task of developing a corporate governance code and implementing an IoD was assigned to the NCCG, which had specifically been created to further corporate governance in the country.

8.2.1.1 Implementation efforts

The Report on Corporate Governance for Mauritius

The NCCG was chaired by Tim Taylor, the then CEO of the Rogers Group; the committee members were selected from private and public sector organisations. Professor Mervyn King, chairman of the South African King Committee, was invited as a consultant to assist with the completion of the project. In 2003, the committee published the report, which incorporated the Code of Corporate Governance and was further revised in April 2004. That same year, the government passed the Financial Reporting Act (FRA) and established the Financial Reporting Council (FRC) and the Mauritius Institute of Professional Accountants (MIPA).

The local Code of Corporate Governance in Mauritius was made applicable to five categories of companies: i) listed companies, ii) banks and financial institutions, iii) large public companies (defined by turnover), iv) state-owned enterprises (SOE), and v) large private companies (NCCG, 2004:15). Based on the 'comply or explain' UK model, these entities were expected to comply with the code by the financial year ending 30 June 2005. In the event of non-compliance, the entity was required to explain *why not*. The code consisted of eight sections: i) compliance and enforcement, ii) boards and directors, iii) board committees, iv) role and function of the company secretary, v) risk management, internal control and internal audit, vi) auditing and accounting, vii) integrated sustainability reporting, and viii) communication and disclosure.

A fundamental element of the local code is its expectation from companies to disclose corporate social responsibility (CSR) activities as part of the integrated sustainability reporting (ISR) and Triple-Bottom Line (TBL) themes. This addition was inspired by the corporate governance developments in South Africa and led by Mervyn King (see Andreasson, 2011). This novel inclusive approach to governance, essential in post-apartheid South Africa, proliferated around the world. Before this, CSR and corporate governance had been deemed separate concepts and researched accordingly. Generally, the stakeholder approach adopted in recent codes has led to the recognition of societal and community aspects in the development of corporate governance (Soobaroyen & Mahadeo, 2008) and has triggered research into the interlinkages between CSR and corporate governance (Bhimani & Soonawalla, 2005; Jamali, Safieddine & Rabbath, 2008).

SOEs play a significant economic, social, and political role in developing countries (OECD, 2015). The NCCG issued Guidance Notes for SOEs on the Code in 2006. The Guidance Notes was comprised of two parts. The first part provides: i) background of the National Code, ii) The Special Circumstances of State-Owned Enterprises, iii) Guidance Notes For State-Owned Enterprises and the second part is the National Code of Corporate Governance.

Mauritius Institute of Directors (MIoD)

The MIoD was launched in August 2007 and officially incorporated in 2008 as a not-for-profit institution under the Companies Act 2001. The MIoD has been instrumental in the development and sustainability of corporate governance in Mauritius. It is important to acknowledge the assistance of the IFC Global Corporate Governance Forum (GCGF) in the formation of the MIoD's interim board. As part of its objective to become a regional leader, the MIoD, together with IoD South Africa, headed the African Corporate Governance Network (ACGN). The official launch of the Executive committee of the ACGN was held in Mauritius in October 2013 with the support of NEPAD Business Foundation; at that time, Jane Valls, the then CEO of the MIoD, was elected as its first chairperson. The MIoD is an affiliate member of the International Corporate Governance Network (ICGN) and a member of the Global Network Director Institutes (GNDI) (www.miod.mu).

8.2.1.2 Selected Surveys on the Code

Mauritius Employers Federation (MEF)

In relation to the CSR requirement, the Code states that companies need to have a code of ethics and are expected to exercise a certain degree of corporate citizenship *vis-à-vis* all its stakeholders, not only its shareholders. In this context, stakeholders are understood to be society, the environment, the employees, the consumers and the local community. According to MEF (2007) study on the level of corporate citizenship of Mauritian enterprises, the following information was obtained; 79% of businesses are involved in internal social initiatives for the benefits of their own employees; such initiatives include training, medical schemes and health and safety standards, which are perceived to be the most common types of CSR initiatives. This finding is compared to 69% of enterprises that are engaged in external social activities such as donations and sponsorships for the benefits of the wider community. In 2010 the Government introduced a CSR legislation mandating profitable companies to contribute 2% of their book profit to government approved Social and Environmental projects. MEF (2011) survey on the implementation of the CSR Levy highlighted the mixed feelings by companies and the report recommended that "it is essential to maintain the flexibility of enterprises to freely manage and spend their CSR funds" (p.14). In MEF (2012) commemorative report these feelings were further expressed as private sector representatives made it clear they were best placed to ensure the proper allocation of CSR resources and investment. Despite several attempts by the Government to introduce guidelines to redress this situation, the CSR levy was repealed in 2015 and a new CSR framework was proposed in 2016.

Mauritius Research Council (MRC)

In 2007, the MRC, the foremost body for promoting and coordinating national investment in research, awarded a grant to the University of Technology, Mauritius under the Unsolicited Research Grant Scheme (UGRS) to undertake a study on the 'Implementation and Impact of Corporate Governance in Mauritius' (Mahadeo & Soobaroyen, 2009). The study had three main objectives: i) investigate the progress and the state of corporate governance implementation in Mauritius by focusing on listed, large non-listed companies and SOEs, ii) examine the impact of corporate governance on the immediate actors and "wealth maximising" stakeholders of the company, and iii) assess the extent and progress of 'integrated sustainability reporting' performed by local organisations. The investigation was a longitudinal survey of companies from 2004 to 2007 and used mixed methods to gather data. The report was submitted in mid-2009, and the results publicly disseminated in November 2009.

The report made the following ten recommendations: i) provide a clearer and stricter wording of the guidelines; ii) clarify the legality and enforceability of the Code; iii) clarify the 'comply or explain' requirement in relation to the wording of the compliance statement and explanations thereof for non-compliance; iv) review the current criteria for large public and private companies (the use of turnover size is convenient but excessively simplistic to allow accurate judgement on corporate governance adoption); v) make annual reports of Statutory Bodies available; vi) establish a separate code of corporate governance and relevant enabling legislation for statutory bodies and all other non-departmental (except Ministries) taxpayer funded bodies, which are not currently subject to the code of corporate governance as a result of their existing legal or regulated status; vii) establish a register of persons able and willing to act as INEDs under the auspices of the MIoD; viii) establish a forum of best corporate governance practice under the auspices of the MIoD; ix) provide greater access to annual reports; and x) extend CSR practices and accountability beyond ad hoc (or structured) charitable donations.

NCCG

In 2009, the NCCG contracted a survey to DCDM, a local accounting firm, to assess the state of compliance with the Code of corporate governance for Mauritius. In all, 318 companies were surveyed; the high level of non-response (41%) prompted an impression of indifference towards corporate governance matters. Of those who responded, only 30% reported that they complied with the code. Expectedly, compliance was high in listed companies (83%) and lower for SOEs (44%). The main findings were the following: i) there is a need to delineate the code from the report, ii) there is no clear understanding regarding how to implement the Code, and iii) the value added of other benefits of the implementation efforts are not apparent (www.nccg.mu).

8.2.2. ROSC 2010

In 2010, the World Bank published an update to the 2002 ROSC. The report was prepared by Berg and Cuttaree. They reported a marked improvement from the previous version. They emphasised the robust legal and institutional framework for corporate governance, which the government and private sector have established. The report also described Mauritius as an international leader in the area of 'board practices and disclosure', and commended the efforts towards CSR disclosures. However, they reported that the ownership structure of large companies has remained concentrated. The core recommendations of the previous ROSC had been applied, but not all of them. Those recommendations that were not implemented were largely related to the Companies Act, which had not yet been amended (ROSC 2010). The Detailed Country Assessment of the OECD Principles registers an evident progress of Mauritius's scores since the last ROSC. The average percent of observance in the shareholder rights chapter has advanced from 60 to 81 and from 60 to 66 in the section on equitable treatment of shareholders. The disclosure percent implementation improved from 60 to 81, and the percent implementation of board responsibilities from 56 to 77. Based on the features of good corporate governance identified by the OECD Principles, the performance of Mauritius is commensurate with leading Asian markets (India, Thailand and Malaysia).

The 2010 ROSC report recommended the following: i) a better anchoring of the code in the legal and regulatory framework and to maintain the continued effort to align the code with the OECD Principles, ii) improvements to the SOE framework, iii) further disclosure of ownership and control, iv) improvements in minority shareholder representation on boards, v) build-up of the rules for related party transactions, and vi) reforms to increase public sector enforcement and build regulatory capacity (ROSC, 2010).

As a solution to improve compliance with the code, the ROSC recommended a move away from the 'comply or explain' to 'comply and explain' principle.

8.2.2.1 Implementation efforts

Changes in legislation

In 2012, the Economic and Financial Measures (Miscellaneous Provisions) Act was enacted with the following amendments to the Financial Reporting Act 2004: all applicable companies must i) submit their annual reports and their reports on corporate governance to the Financial Reporting Council, and ii) submit a 'Statement of Compliance' that 'would determine the extent of compliance with the requirements of the Code'.

The Statutory Bodies (Accounts and Audit) Act 2009 was amended to make it obligatory for statutory bodies to incorporate a corporate governance report in their annual reports that is in accordance with the National Code of Corporate Governance (NCCG).

The National Code of Corporate Governance for Mauritius

In February 2017, the NCCG issued the 'New Code of corporate governance' partly based on the recommendations of the ROSC 2010. The New Code adopts the recommended 'apply and explain' principle, which is based on the philosophy of application coupled with disclosure. The code is based on the following eight principles: i) Governance Structure, ii) The Structure of the Board and its committees, iii) Director Appointment Procedures, iv) Director Duties, Remuneration and Performance, v) Risk Governance and Internal control, vi) Reporting with Integrity, vii) Audit, and viii) Relations with Shareholders and Other Key Stakeholders (www.nccg.org).

The New Code of corporate governance is meant specifically for all public interest entities defined in the FRA 2004 and public sector organisations listed as public interest entities in Schedule 1 of the FRA 2004 and are expected to apply the New Code by the end of the financial year ending 30th June 2018. Other companies are encouraged to duly consider the application of the code insofar as the principles are applicable. Moreover, the code provides direction for the following: i) Auditors to report on the compliance with the Code of Corporate Governance as per International Standards on Auditing (ISA) 700 and ii) the statement to be provided to the Financial Reporting Council. The New Code also encourages companies that do not previously have a Code of Ethics to develop one and in keeping with international trends, the code has added 'gender' as a component of board diversity (www.nccg.org).

The code provides additional application guidance to the following organisations: i) banks, ii) companies listed on the SEM, iii) unlisted family companies, iv) statutory bodies, v) groups and their subsidiaries and vi) Category 1 Global Business Licence and Management companies.

8.2.2.2 Local Corporate Environment: internal and external

The framework developed by the World Bank Group (1999) has been used to depict the main players in the internal and external environments in which companies in Mauritius operate (See Appendix). The adoption of the codes of corporate governance has necessitated significant changes in the regulatory and legal corporate environments. The framework lays emphasis on the country's legal, regulatory, professional and business environments. It is important to note that prior to finalising both codes of corporate governance, extensive effort was made by the NCCG and the Ministry of Finance and Economic Development to engage in a public consultation.

8.2.2.3 Regional and International Support

Private-sector-funded institutions such as the Commonwealth Association of Corporate Governance (CACG) and other interested parties (e.g., consulting firms and business schools) provide for the corporate governance agenda towards educating, consulting and informing corporate governance in developing countries. Corporate governance obtained further significance on the African continent, following its inclusion in the African Peer-Review Mechanism (APRM). The APRM is an African-led initiative that provides member countries with the possibility to self-monitor all aspects of their governance and socio-economic development. Mauritius had the opportunity of being reviewed by the APRM. The findings were published in 2010 and were not much different from what the 2010 ROSC reported. The main recommendations were as follows: i) Mauritian Authorities must improve public sector governance and reporting, ii) address basic shareholder rights under the Companies Act and iii) additional protection for minority shareholders' rights (APRM, 2010).

8.3 Corporate Governance Developments

8.3.1 Adoption of the Code by local enterprises

Following the enactment of the first code of corporate governance in Mauritius, a marked improvement in the implementation of the code was observed for listed companies. The improvement started with a sharp rise from 2004 and 2005 (Figure 1) which may be because adherence to the code is a listing requirement and by 2007, the increase had levelled off below expectations (Mahadeo & Soobaroyen, 2016). Whilst corporate governance scores have increased in general, the discernible changes were the board composition, audit and corporate governance committee. The study also reveals a possible partial association between labour intensiveness and corporate governance implementation. Another interesting statistical result was observed: a strong and consistent relationship between the various corporate governance scores and Independent Non Executive Director (INED) representation amongst listed companies, this points to the critical role of the INED as a 'change agent' for those companies which took the decision to implement this aspect of the corporate governance code (Mahadeo & Soobaroyen, 2016). This is additional evidence that the stakeholder inclination of the code has been met (Ararat & Dallas, 2011; Soobaroyen & Mahadeo, 2012).

However, the situation of listed enterprises needs to be contrasted to a fairly low level of interest in the implementation of the code and the resulting low implementation scores among LPP companies as shown in Figure 1 (Mahadeo & Soobaroyen, 2009). The progress of LPPs from 2004 to 2007 has been much slower than envisioned. Figure 1 shows that listed companies have implemented the code almost three times faster than the LPPs. A closer examination of the annual reports of LPPs and interview findings revealed that some companies chose not to implement any aspect of the Code while others reached fairly high scores of implementation.

FIGURE 1 - Total Corporate Governance Score per category of company 2004-2007 (max score 146)

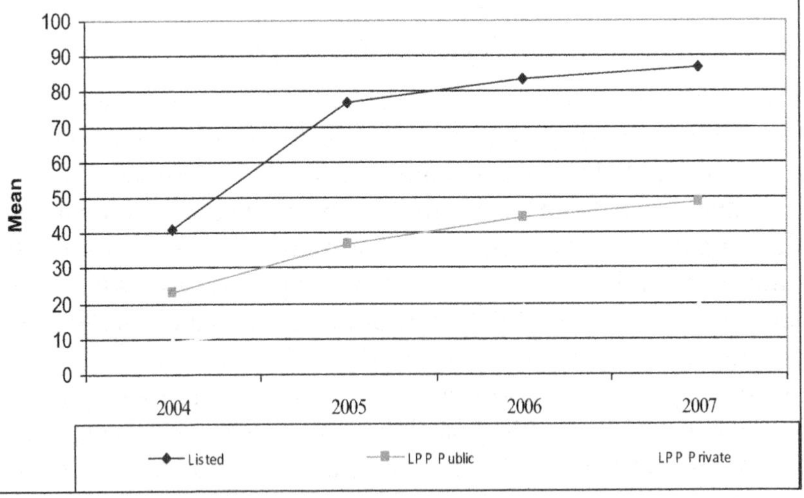

Source: Mahadeo & Soobaroyen (2009)

The implementation of the Code among SOEs has been much more problematic due to key differences between SOEs and publicly traded corporations (Bozec, Zeghal & Boujenoui, 2004). Mahadeo & Soobaroyen (2012) revealed the various impediments to the applicability of the code to SOEs namely: i) definitional issues of the term SOE, ii) the blurred nature of ownership and accountability in the sector, iii) the almost non-existent financial accountability and iv) existing political rivalries. As mentioned earlier, in 2006 the NCCG issued a separate 'Note of Guidance for SOEs' to suit their realities. The investigation of Mahadeo & Soobaroyen (2012) however found little progress in the quality of corporate governance disclosures except for rare cases where the SOE seemed to have engaged in lip service of corporate governance concept. It is noteworthy, that there has been negligible evidence of SOE's performance except through extracts of the National Audit Office's reports. For instance, the Code highlights an important section on the clear division of responsibilities between the chairman and the CEO. However, the 2005-2006 Report of the Director of Audit noted the poor performance in many of the government enterprises as the result of a conflicting relationship between the CEO and chairperson.

8.3.2. Board Practices

ROSC 2010 has laid emphasis on the robust legal and institutional framework for corporate governance, which the government and private sector have set up. The report qualified Mauritius as an international leader in the area of 'board practices and disclosure' amongst others. This is confirmed by the qualitative study of Soobaroyen and Mahadeo (2012), which based on Roberts' (2001) typology of governance systems reported a move from a sovereign governance stance, which is typified by low involvements of non-executive directors, poor disclosure and transparency and supremacy of executive directors in the management of companies to an individualising form of governance. The individualising form of accountability is evidenced by a marked improvement of directors' responsibilities towards shareholders, by the belief that corporate governance is mainly about disclosure and transparency rather than external inspection and the use of board sub-committees to assist and endorse managerial moves. Moreover, the authors also compared the results with the Pettigrew and McNulty's (1995) notion of maximalist and minimalist boards. It was evident from the data that Mauritian companies had traditionally adopted minimalist boards and a gradual change has started towards a maximalist board, with the accompanying change towards board responsibility, composition and committees,. This situation resonates with the findings of Baysinger and Hoskisson (1990) that board composition ultimately influences directors' ability to provide strategic direction and performance.

8.3.3 Ownership Structure

The 2002 ROSC laid emphasis on the fact that the economy of the country was dominated by a small group of family-owned companies and only 25% of shares were owned by non-family members in listed companies. Goldsmith (2007) who also observed this instance indicated that good Corporate Governance was lacking during the country's 'miracle' years. This situation represented important challenges to promote corporate governance. Despite being listed on the SEM, those companies were still controlled by family holding company or a partnership acting as holding company. Those holding companies controlled diverse enterprises and owned vast land holdings, which have failed to produce satisfactory earnings. As a result, 'unlocking shareholder value' was identified as a fundamental issue for the Mauritian economy. The report 2002 ROSC made 'unlocking shareholder value' its first recommendation with a 'high' priority label.

Notwithstanding the steps taken by the State to unlock shareholder value, the 2010 ROSC again commented on the high concentration of the ownership in the large companies which has persisted with almost 60% of companies having one controlling shareholder (holding more than 20% of shares) as at financial year end 2010 (SEM, 2011). This situation is akin to the Type II Agency Problem (Villalonga & Amit, 2006) prevalent in Continental Europe and Asia where family businesses are predominant. This, therefore, largely questions the effectiveness of the recommended measures by the World Bank of utilising market for corporate control (Becht & Roell, 1999; La Porta, R., Lopez-de-Silanes, F., & Shleifer, 1999). This is of significant interest since a number of similar economies have been encouraged to adopt Western-oriented corporate governance mechanisms, even though it is arguable that such mechanisms match their national contexts (Pillay, 2015; Sobhan, 2016).

In order to make sense of this situation, the Bebchuck and Roes' (1999) theory of path dependence which argues that the corporate structure which an economy possesses at any point in time is partly dependent on those it had at earlier times will be employed. The Bebchuck and Roe theory informs as to why enterprises in some economies even when pressured to converge their corporate governance, still vary in their ownership structure. They bring about two paths: structure and rule driven. When applying the pattern in ownership concentration of most listed companies (SEM, 2017) to the Bebchuck and Roe's theory, it appears that the case of Mauritius is closely related to '*Persistence of Existing Structures due to Rent Seeking*'. Predominantly, the phenomenon is an instance of '*persistence of concentrated ownership*' and is due to '*the controller's roadblock*'. Schmidt and Spindler (2002) concur with Bebchuck and Roe's proposition that expecting rapid conversion from one ownership structure to a proposed one is challenging. Bebchuck and Roe recommend two possible solutions to curb persistence of concentrated ownership which could be considered by local policymakers: i) application of the Coase Theorem, and ii) limiting persistence: large inefficiencies or rent destroying rules.

8.3.4 Foreign Direct Investment (FDI)

The efforts of the Government to attract investment paid off rapidly and by 2007 the country was declared an attractive destination to undertake business by international bodies such as the World Bank. According to figures published by the latter, FDI reached its highest peak in 2012 by gaining from significant investment totaling MUR (million) 20373 (Figure 2 below). The Bank of Mauritius commented on this exceptional performance despite the prevailing Eurozone crisis and global recession. Mauritius has topped the 'ease of doing business' ranking in Africa and surpassed South Africa. In 2014, the IMF Article IV consultation advised the Mauritian Government to 'consolidate recent macroeconomic gains, strengthen policy buffers, and pursue greater economic diversification through structural reforms to enhance resilience of the economy' due to uncertain external environment which holds risks (IMF, 2014). The World Economic Forum also commented on the overall good performance of the Mauritian Economy as compared to other African nations and was exemplified by the country's securing the top spot in Sub-Saharan Africa in the annual Global Competitiveness Report for 2016-2017 (WEF, 2017).

Fig. 2: Foreign Direct Investment Inflows (in MUR Million) from 2006 to 2015

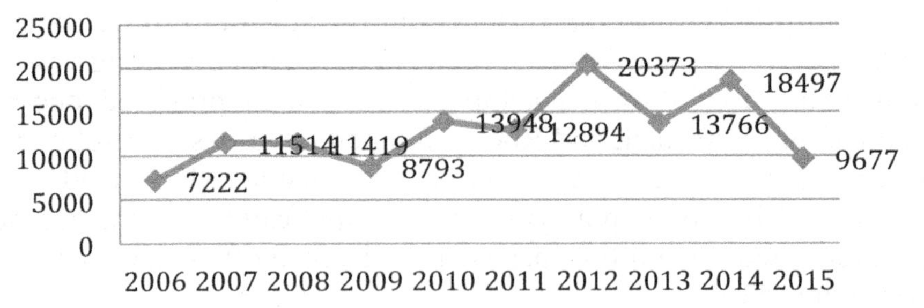

Source: (adapted from the World Bank, 2017)

8.3.5 Integrated Sustainability Reporting (ISR)

As mentioned earlier, the corporate governance code in Mauritius has adopted, the stakeholder approach under the label ISR (NCCG, 2004). This has given researchers (Mahadeo, Oogarah-Hanuman, and Soobaroyen, 2011; Soobaroyen & Mahadeo, 2016) an interesting rationale to assess the impact of corporate governance code from this perspective and with regard to the theoretical concept that would exemplify CSR activity in Mauritius. The classic study of Rossow (2005) on the African continent has created the expectation that corporate governance developments will contribute to better societal change (through enhanced accountability and ethical conduct) in developing nations. The local code of corporate governance (2004) showed a willingness to demonstrate the specific societal issues the country has to deal with and identified the following four categories: i) ethics, ii) social, iii) environment and iv) health and safety. Using content analysis of annual reports of listed companies, Mahadeo *et al* (2011) analysed the four categories, by lumping them under the banner Social Environmental Reporting (SER). The study found noteworthy yet selective increase in volume (of words) and quality of SER from 2004 to 2007. Based on Suchman's (1995) conceptualizations of legitimacy, the research contends that the changes were related to the necessity felt by companies to establish their affiliation to pro-social objectives (i.e moral legitimacy) and, to a much lesser degree, were inspired by the need to manage specific stakeholders (i.e pragmatic legitimacy). When considered in isolation, ethical disclosures have increased which is reflective of the endeavour to attain procedural legitimacy as an answer to criticisms related to corruption and unfair/unethical business processes. Figure 3 below shows the disclosure trend of the four categories independently and the aggregated score. Moreover, the MIoD launched its Ethics Guide for Boards in 2013 to assist directors in implementing their legal responsibilities in a just fashion (Schwartz, Dunfee & Kilne, 2005).

FIGURE 3: Weighted Overall and Detailed CSR Scores for Listed Companies (2004-2007)

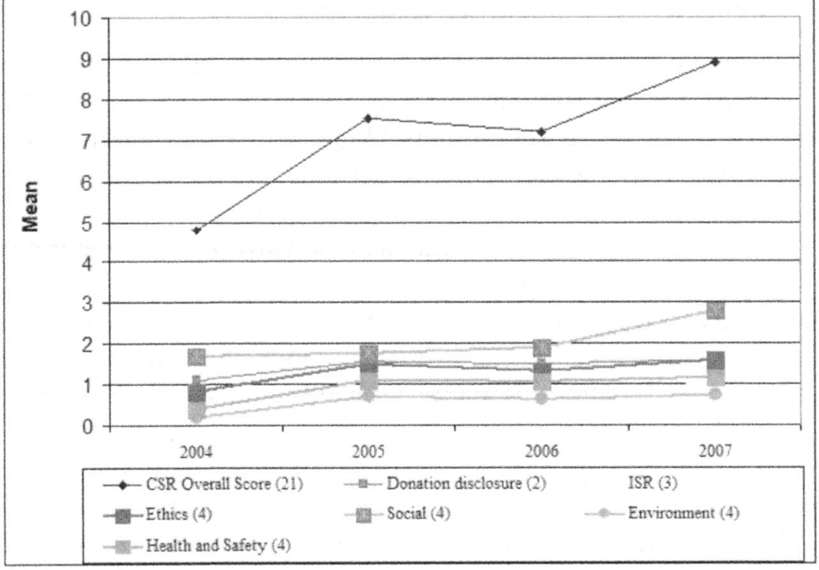

Source: Mahadeo & Soobaroyen (2009)

8.4 Conclusion

Based on academic investigations, local, regional and international surveys of corporate governance developments, reforms and achievements in Mauritius, it is evident that the steps that the country took to achieve global standards appear to have worked fairly well. The latest ROSC has commended the country as an international leader in 'board practices and disclosures'. The country's advancements in board processes were amply supported by research. Moreover, the nation's efforts to attract FDI were certainly the appropriate ones and rapidly produced the expected benefits. By 2007, the island was confirmed in its position as an attractive destination for FDI (OECD/ADB). Essentially, these reforms have also contributed to shielding the country from the shocks of the global recession and Eurozone crisis. Yet local policy makers still have work to do to reach the recommended and desired 'unlocking of shareholder value' and to protect minority shareholders.

Then again, the recurring local corporate scandals have ensured that Corporate Governance remains high on the government's agenda. New laws have been enacted and existing laws have been amended to ensure improved protection of all stakeholders. A positive trend towards the adoption of international corporate governance norms is manifest as the monitoring roles of local institutions have sharpened.

In summary, it is important to appreciate that corporate governance is essentially an evolving organizational process which takes time. It is only reasonable to anticipate a long evolving implementation process instead of a rapid drastic change to ensure that corporate governance develops into a reliable means for development.

REFERENCES

Andreasson, S. (2011). Africa's prospects and South Africa's leadership potential in the emerging markets century. *Third World Quarterly*, *32*(6), 1165-1181.

Alladin, I. M. (1993). *Economic Miracle in the Indian Ocean: Can Mauritius Show the Way?* Editions de l'océan Indien.

APRM (2010), "APRM country review report no. 13: the Republic of Mauritius", available at: http://aprm-au.org/view-publication?nxtpbdi=10&&pubId=170

Ararat, M, and Dallas, G., (2011) "Corporate governance in emerging markets: Why it matters to investors-And what they can do about it." *Private Sector Opinion, Issue 22, International Financial Corporation (IFC) Global Corporate Governance Forum*. Retrieved from

http://ssrn.com/abstract=1914267

Baysinger, B., & Hoskisson, R. E. (1990). The composition of boards of directors and strategic control: Effects on corporate strategy. *Academy of Management review, 15*(1), 72-87.

Bebchuk, L. A., & Roe, M. J. (1999). A theory of path dependence in corporate ownership and governance. *Stanford law review*, 127-170.

Becht, M, & Roel A. (1999). Blockholding in Europe: An International Comparison. *European Economic Review*, 43 1049-1056

Bhimani, A., & Soonawalla, K. (2005). From conformance to performance: The corporate responsibilities continuum. *Journal of Accounting and Public Policy, 24*(3), 165-174.

Brautigam, D., & Diolle, T. (2009). Coalitions, capitalists and credibility: overcoming the crisis of confidence at independence in Mauritius. *DLP Research Paper, 4.* Available at: www.lpdlec.org.

Braun, V., & Clarke, V. (2006). Using thematic analysis in psychology. *Qualitative research in psychology, 3*(2), 77-101.

Bozec, R., Zéghal, D., & Boujenoui, A. (2004). The Effect of the Reform of Canadian State-Owned Enterprises on Major Corporate Governance Mechanisms. *Australian Journal of Public Administration, 63*(2), 79-94.

Coleman, B., & Aalund, D. (1998). Deutsche Bank plan may shake Germany. *The Wall Street Journal Europe, 16*, 13.

Gibson, M. S. (2000). Big Bang deregulation and Japanese corporate governance: a survey of the issues. In *Crisis and Change in the Japanese Financial System* (pp. 291-314). Springer US.

Goldsmith, A. A. (2007). Is governance reform a catalyst for development?. *Governance, 20*(2), 165-186.

Houbert, J. (1981). Mauritius: independence and dependence. *The Journal of Modern African Studies, 19*(1), 75-105.

Ibrahim Foundation (2016). The 2016 Report Ibrahim Index of African Governance: Key Findings. Retrieved from: http://mo.ibrahim.foundation/news/2016/progress-african-governance-last-decade-held-back-deterioration-safety-rule-law/

Jamali, D., Safieddine, A. M., & Rabbath, M. (2008). Corporate governance and corporate social responsibility synergies and interrelationships. *Corporate Governance: An International Review*, *16*(5), 443-459.

KPMG Advisory (2005). Corporate Governance Practices Survey 2005/2006, KMPG Internal Audit Services, Mauritius.

Krippendorff, K. (2004). *Content analysis: An introduction to its methodology*. Sage.

Lange, M. (2003). Embedding the Colonial State. *Social Science History*, *27*(03), 397-423.

La Porta, R., Lopez-de-Silanes, F., & Shleifer, A. (1999). Corporate ownership around the world. *The journal of finance*, *54*(2), 471-517.

Mahadeo, J. D & Soobaroyen, T. (2009). A Study of the Implementation and Impact of Corporate Governance in Mauritius, Mauritius Research Council Grant Report: MRC/RUN-A061/16

Mahadeo, J. D., Oogarah-Hanuman, V., & Soobaroyen, T. (2011, September). Changes in social and environmental reporting practices in an emerging economy (2004–2007): Exploring the relevance of stakeholder and legitimacy theories. In *Accounting Forum* (Vol. 35, No. 3, pp. 158-175). Elsevier.

Mahadeo, J. D., & Soobaroyen, T. (2012). Corporate governance implementation in an African emerging economy: The case of state-owned entities. In *Accounting in Africa* (pp. 227-254). Emerald Group Publishing Limited.

Mahadeo, J. D., & Soobaroyen, T. (2016). A longitudinal study of the implementation of the corporate governance code in a developing country: the case of Mauritius. *Business & Society*, *55*(5), 738-777.

Meade, J. E. (1961). Mauritius: A case study in Malthusian economics. *The Economic Journal*, *71*(283), 521-534.

Mauritius Employers Federation. (2007). Are Mauritian Enterprises Good Corporate Citizens? The Role of Business in society, Economics Department, Mauritius Employers Federation.

Mauritius Employers Federation. (2011). Survey Report on the Practical Implementation of CSR Under the New Legislation, Mauritius Employers Federation.

Mauritius Employers Federation. (2012). Celebrating 50 Years: Commemorative Report 2012, Mauritius Employers Federation.

Meisenhelder, T. (1997). The developmental state in Mauritius. *The Journal of Modern African Studies*, *35*(02), 279-297.

Miles, W. F., (1999). The Mauritius Enigma. *Journal of Democracy*, 10,2:91-104.

OECD (2015). *OECD Guidelines on Corporate Governance of State-Owned Enterprises, 2015 Edition*, OECD Publishing, Paris.

Pettigrew, A., & McNulty, T. (1995). Power and influence in and around the boardroom. *Human relations*, *48*(8), 845-873.

Pillay, R. (2015). *The Changing Nature of Corporate Social Responsibility: CSR and Development–The Case of Mauritius*. Routledge.

Roberts, J. (2001). Trust and control in Anglo-American systems of corporate governance: The individualizing and socializing effects of processes of accountability. *Human Relations*, *54*(12), 1547-1572.

Rossouw, G.J. (2005), Business Ethics and Corporate Governance in Africa, *Business and Society*, 44(1), 94-106.

Schmidt, R. H., & Spindler, G. (2002). Path dependence, corporate governance and complementarity. *International Finance*, *5*(3), 311-333.

Schwartz, M. S., Dunfee, T. W., & Kline, M. J. (2005). Tone at the top: An ethics code for directors? *Journal of Business Ethics*, *58*(1-3), 79.

Sobhan, A. (2016). Where institutional logics of corporate governance collide: Overstatement of compliance in a developing country, Bangladesh. *Corporate Governance: An International Review*, *24*(6), 599-618.

Soobaroyen, T., & Mahadeo, J. D., (2008). Selective compliance with the corporate governance code in Mauritius: Is legitimacy theory at work?. In *Corporate Governance in Less Developed and Emerging Economies* (pp. 239-272). Emerald Group Publishing Limited.

Soobaroyen, T., & Mahadeo, J.D., (2012). Do corporate governance codes improve board accountability? Evidence from an emerging economy. *Qualitative Research in Accounting & Management*, *9*(4), 337-362.

Soobaroyen, T., & Mahadeo, J. D. (2016). Community disclosures in a developing country: insights from a neo-pluralist perspective. *Accounting, Auditing & Accountability Journal*, *29*(3), 452-482.

Suchman, M. C. (1995). Managing legitimacy: Strategic and institutional approaches. *Academy of Management Review*, *20*(3), 571-610.

The World Bank (2017) Doing Business 2017: Equal Opportunity for All Retrieved from:

http://www.doingbusiness.org/reports/global-reports/doing-business-2017

The World Bank and IMF. 2002. Report on the Observance of Standards and Codes: Corporate Governance Country Assessment Mauritius, Ministry of Economic Development, Financial Services and Corporate Affairs.

The World Blank and IMF. 2010. Report on the Observance of Standards and Codes: Corporate Governance Country Assessment Mauritius, Ministry of Economic Development, Financial Services and Corporate Affairs.

The World Bank (2017) FDI Net Inflows available at: http://data.worldbank.org/indicator/BX.KLT.DINV.CD.WD

United Nations Development Programme. (2007). Building the Capable state in Africa: Mauritius Response Paper, Oagadougou, Burkina Faso.

Villalonga, B., & Amit, R. (2006). How do family ownership, control and management affect firm value? *Journal of Financial Economics*, *80*(2), 385-417.

Wilmshurst, T. D., & Frost, G. R. (2000). Corporate environmental reporting: a test of legitimacy theory. *Accounting, Auditing & Accountability Journal*, *13*(1), 10-26.

World Economic Forum (2017) The Global Competitiveness Report 2016-2017. Retrieved from:

http://www3.weforum.org/docs/GCR2016-2017/05FullReport/TheGlobalCompetitivenessReport2016-2017_FINAL.pdf

Appendix

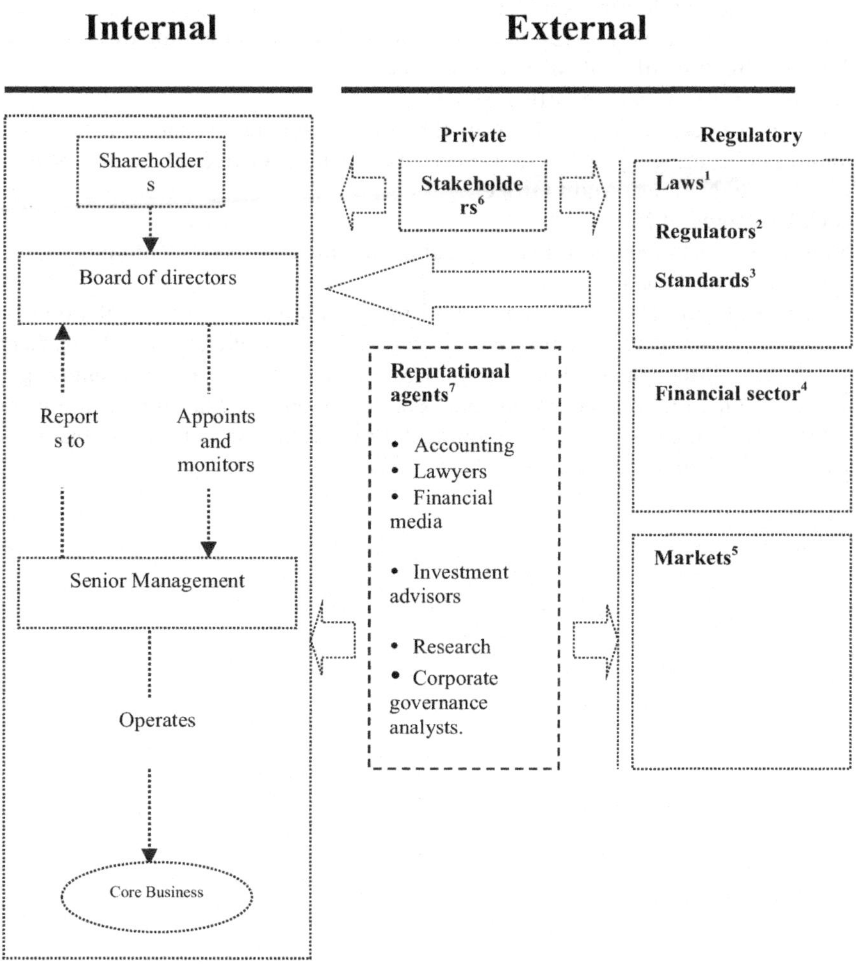

Figure A: Corporate Governance Environment

(Source: World Bank Group Report, 1999)

End Notes

[1] Companies Act 2001, Financial Reporting Act, Financial Services Act 2007, Statutory Bodies Act, (CSR Levy 2010 repealed in 2015), The Good Governance and Integrity Act 2015

[2] National Committee on Corporate Governance, Mauritius Institute of Directors, Mauritius Institute of Professional Accountants

[3] International Financial Reporting Standards

[4] Financial Reporting Council, Registrar of Companies, Financial Services Commission, Integrity Reporting Services Agency, Office of Public Sector Governance, National Audit Office

[5] Stock Exchange of Mauritius

[6] Employees, Customers, Suppliers, Local Community, National Community, the Government of the Day

[7] International and Local Accounting Firms, Commonwealth Association for Corporate Governance, World Bank, International Monetary Fund, African Peer Review Mechanism, University of Mauritius, University of Technology, Mauritius, Open University, Mauritius Research Council, Mauritius Employers Federation, United Nations Development Programme, stockbrokers, business analysts (local and international)

Chapter 9

Corporate Governance in Nigeria

Elewechi Okike and Peace Okougbo

9.0 Introduction

With a population of over 195 million[87], comprising 250 ethnic groups, and a landmass of 924,000sq.km, Nigeria is the third largest country in the Commonwealth after India and Pakistan, but the most populous in Africa. Nigeria joined the Commonwealth in 1960 and produced the third Secretary General, Chief Emeka Anyaoku.

9.1 Historical Background

The colonisation of Nigeria by the British started with the conquest of Lagos in 1851 and further conquests of other strategic towns in the hinterland until the British control over the area now known as Nigeria was ratified at the Berlin Conference in 1885, where the concept of the 'spheres of influence' was developed (Wallace, 1989). The British stayed in Nigeria until 1960 when by mutual agreement between the Nigerian political activists seeking self-government for the country and Her Majesty's Government, the country became independent and a member of the British Commonwealth of Nations (Wallace, 1989).

9.2 Country Structure and Government

The country is a federal republic of 36 states including the federal capital territory, Abuja. Since Nigeria became an independent country on 1 October 1960, the economy has been shaped by the experimentation with different types/styles of leadership and government (parliamentary democracy, UK style; presidential democracy, US style; and military dictatorships of different kinds (Okike, 1994, 2004). After the first military coup in 1966, and the subsequent civil war in 1967-1970, the army remained the dominant political player, except between 1979-1983 and for a few weeks in 1993. The sudden death of the dictatorial military ruler, Sani Abacha, led his successor to move quickly to restore human rights, repair relations with the West and install a democratically elected government in 1999. Under the new constitution adopted in May 1999, a strong executive presidency appoints a Federal Executive Council, comprising government ministers and ministers of state from each of Nigeria's 36 states. Although the 36 states enjoy greater autonomy than under the military administration, they remain dependent on the central government for funding. The executive is accountable to the bicameral National Assembly, though in practice personal and ethnic ties dominate the political process (Okike, 2004). Nigeria's geo-political boundaries, a creation of the British colonial government, were arbitrarily imposed, cutting across ethnic groups and cultural units and paying little regard to the people's history and past experiences (Barbour et al., 1982). As a result, since gaining independence, Nigeria has grappled with the problems and crises deeply rooted in tribalism and other ethnic tensions. These tensions have led to successive military coups, including tribal and religious unrests. In addition, the country has had a long history of gross mismanagement of the economy and corruption at all levels of government (Okike, 2004).

9.3 Accountability and Corruption

Undeniably, accountability in all facets of the economy is an essential ingredient to the economic development of any nation. But various governments in Nigeria have been contending with the problem of achieving adequate accountability in the country (Okike, 2004). This is because Nigeria is a country of many paradoxes. On the one hand, it is deemed to be of economic relevance globally because it is one of the world's largest producers of oil. This equally makes the condition of the Nigerian business environment of great importance. On the other hand, it has been ranked as one of the poorest nations in the world, being the 5th country with a high percentage (70 per cent) of its population living below the poverty line[88]. Furthermore, Nigeria has secured for itself a reputation as a country with a pervasive state of corruption in both private and public spheres (Okike, 2004). Over the years, the state of corruption in Nigeria has reached epic proportions (Transparency International, 2017). The latest figures released by Transparency International shows that the Corruption Perception Index (CPI) of Nigeria is still high, 148 out of 180 countries, with a score of 27 out of 100[89]. This lends credence to why corporate governance remains high on the agenda in Nigeria both as a developing country and as a member of the Commonwealth.

9.4 Background to Company Regulation

Issues relating to the regulation, control and governance of business enterprises in Nigeria are largely contained within the provisions of company legislation. This system of legislation has its roots in Nigeria's colonial past (Okike, 2007). Like most former colonies of Britain, Nigeria 'inherited' its legal and corporate structures from the UK. However, before the advent of the British Government (in Lagos in 1861) and in Nigeria as a whole in 1914, European traders, who were mostly Portugese traded with local people along the coasts of West Africa and kept records of their transactions for the mutual benefit of both parties (Anibaba, 1990). This established some form of accountability between the parties. But during the colonial period, British accounting was introduced into the country and Nigeria's legal system mirrored the UK pattern (see Parker, 1958). At independence, the Companies Ordinance of 1922 (based on the UK act of 1908) became Nigeria's Companies Law, which was repealed and replaced by the Companies Act of 1968. The Act was a marked improvement on the previous law, introducing for instance, mandatory provisions for accounts and encouraged greater accountability of directors and more effective participation of shareholders in the affairs of the company (Okike, 1994). Despite attempts to assert Nigeria's political and economic independence, the 1968 Act mirrored the UK Companies Act of 1948 to a large extent (Okike, 2007). This was because at that time, foreigners, mostly British, controlled the activities of business enterprises and along with their economic interests brought with them their company legislation. The consequence of simply mimicking British company legislation was that it failed to deal with company law problems that were peculiar to Nigeria's socio-cultural and political environment. It also did not address the rapid economic and commercial developments of the country (Okike, 2007).

Another defect of the Companies Act 1968 was its failure to communicate with Nigerians in a simple and comprehensible manner. The need for reform of the law became inevitable, more so, with the introduction of the Nigerian Enterprises Promotion Acts, which were intended to promote the indigenisation exercises. The implementation of the provision of these Acts revealed the many inadequacies of the Companies Law and it was clear that a more relevant set of principles and legal practice was required. These concerns impelled the Government through the Attorney General and Minister of Justice to direct the Nigerian Law Reform Commission in 1987 to undertake a review and reform of Nigerian Company Law[90]. The Committee submitted its report together with three draft decrees, which the Government accepted and immediately referred to a National Consultative Assembly on Company law for deliberation and recommendation for further action. The Assembly comprised leaders of thought and persons with informed and working knowledge of company law and practice including lawyers and accountants.

The Consultative Assembly's Report including a draft decree was submitted after one year's deliberation. Following this, the government promulgated the Companies and Allied Matters Decree 1990 (No.1 of 1990) (CAMD, 1990) with effect from the 2nd of January 1990. Several amendments were made to the Decree over the years, resulting in the Companies and Allied Matters Act, Cap.C20, Laws of the Federation of Nigeria 2004. By this time, the country had ceased to be under military rule, and was being run by a democratically elected government.

Before highlighting the curious provisions of CAMD No.1 1990, it is worth mentioning that the final legal document which arose from the dual and consecutive deliberations of the Law Reform Commission and the Constituent Assembly was a carbon copy of the UK Companies Act 1985. The only extraneous additions were the curious and contentious ones. Okike (1994) provides an interesting insight into these curious provisions, which were an attempt to reflect Nigeria's peculiar socio-economic and political culture. These peculiar provisions included those relating to the appointment of auditors, auditors' reports, the services that auditors can perform and the potential threat to the monopoly of the marketing for auditing services by members of the Institute of Chartered Accountants of Nigeria (ICAN). On the auditor's report, a peculiarly Nigerian innovation was the requirement that the report should be countersigned by a legal practitioner (sec.359(2))[91]. Whereas members of the accounting profession saw this latter provision as an embarrassment, the lawyers regarded it as a step in the right direction.

Auditors, legal practitioners and the general public received the curious provisions of the CAMD 1990 with mixed feelings. Unsurprisingly, the 'offending' provision of section (359(2)) was repealed, after a year of experimentation with it. The Decree was severely criticised on the grounds that it was evasive over a number of issues in relation to the implementation of the provision of sec.359(2). The Decree did not state what role the legal practitioner was expected to play in the actual examination of the company's books and records; nor did it state which legal practitioner should countersign the auditor's report – the client company's or the auditor's. In addition, by the provisions of sec.368(2) the legal practitioner is absolved from legal liability for negligence in the issuance of the auditor's report. For a detailed analysis of the experimentation with the provisions of sec.359(2), see Okike (1994), and Okike (2004), for how the accounting profession responded to the challenge to its legitimacy.

9.5 The Current Regulatory Framework for Companies

A country's regulatory framework is critical in determining the buoyancy of the business environment and its ability to attract foreign investment. Thus the Nigerian Senate recognises that the legal framework for undertaking business in Nigeria must reflect the challenges of modern markets in which business and investment decisions are increasingly determined by global conditions. This increasingly global market place should incorporate regulatory conditions, which are modelled around international best practice. CAMA is one of the most critical pieces of legislation, which impacts the Nigerian business climate and Micro, Small and Medium Scale Enterprises (MSMEs). It also directly affects the influx of Foreign Direct Investment (FDI) into Nigeria because of its relevance to ease of doing business and ease of investing in Nigeria[92]. Appropriately, the Nigerian Senate on the 15th day of May 2018 passed the Companies and Allied Matters Act (CAMA), 2004 (Repeal and Re-enactment) Bill, 2018 ("the Bill") into law. The Bill, which seeks to repeal and re-enact the existing Companies and Allied Matters Act 2004, is aimed at addressing the shortcomings of CAMA 2004, and when assented to by the President, is expected to promote the ease of doing business in

Nigeria. The Senate considered that the current legislative framework was now more than 20 years in existence with minimal amendments during the period. The consequence was that the corporate landscape was not in tandem with developments in the global market place, and there was the need to make necessary amendments to make the Act more contemporary and relevant.

The Bill is divided into eight (8) parts as follows:

Part A – deals with the administrative aspect of the Corporate Affairs Commission
Part B – deals with the incorporation of companies in Nigeria and incidental matters
Part C – deals with Limited Liability Partnership
Part D – deals with Limited Partnership
Part E – deals with Business Names
Part F – deals with Incorporated Trustees (non-profit organisations)
Part G – deals with establishment of Administrative Proceedings Committee
Part H – provides the Short Title of the Bill

The summary of the repeal and re-enactment of CAMA:

- *Single Member Companies* – a single person can form a company for the first time in Nigeria. This provision is consistent with what obtains in other advanced countries such as the United Kingdom and Singapore
- *Limited Liability Partnerships* – Limited Liability partnerships can now be formed
- *Financial Assistance* – companies will now be permitted to provide financial assistance to their shareholders under the new bill
- *Reduction in Share Capital* – private companies can reduce their share capital if a special resolution to that effect is

passed – without the added burden of applying to court for a confirmation of the reduction.

- *Resolving Insolvency* – CAMA introduces a company rescue and insolvency legal regime to save viable businesses and to ensure that non-viable businesses can exit the market quickly, allowing the deployment of assets to more productive firms.
- *Company Secretary* – the provisions ease the regulatory burden on public companies to appoint a Company Secretary, and make it optional for small companies with one shareholder.
- *Annual General Meeting* – it would no longer be mandatory for small companies to hold AGMs
- *Minority Shareholder Rights* – this is geared towards enhancing minority shareholder rights and protects the rights in corporate governance
- *Beneficial Ownership* – mandates the disclosure of beneficial interests in a company's shares and prescribes punitive measures for failing to disclose such interests.
- *Exemption from Audit* – small companies are exempted from appointing auditors if (i) it has not carried on business since its incorporation; or in a particular financial year; and (ii) where the company's turnover is not more than N10m and its balance sheet total is not more than N5m.

The Senate believes that the introduction of these new provisions in the act will greatly improve ease of doing business and generally enhance the business climate in Nigeria, and would have the following economic impact[93]:

i. *Ensure more business-friendly regulation for Micro, Small and Medium Enterprises*
ii. Fewer reporting obligations for small companies

iii. Reduction in time and cost for setting up a company
iv. Promotion of financial stability
v. Increasing investor confidence in the Nigerian financial sector as well as all sectors of the economy

These proposed amendments to the Act would have the overall effect of making Nigerian Company Law more fit for today's business realities, improve the business environment and performance across the Nigerian economy as a whole, as well as reduce direct compliance costs for businesses in Nigeria[94].

9.6 The System of Corporate Governance in Nigeria

The system of corporate governance existing in any country is often shaped by a wide array of internal as well as external factors (Okike, 1994, 1998, 1999; Okike & Adegbite, 2012). Internal factors include the state of maturity of the economy and the capital market, corporate and business cultures, the legal system, government policies and vibrant professional and regulatory bodies, amongst others.

Given Nigeria's colonial heritage, the system of corporate governance is essentially "Anglo-Saxon, or the "outsider control system" (Franks and Meyer, 1994). Features of this system are as follows (Okike, 2007):

- The interest of shareholders is paramount in the day-to-day activities of management, and their priority is the maximisation of shareholders' wealth
- There is a functioning capital market, which helps to align the interests of management and shareholders, through right to buy and sell shares at prices which reflect their value as perceived by investors.

- There is a chain of accountability. Company executives are accountable to the board of directors, who are in turn accountable to shareholders.
- The rights and responsibilities of key players in the corporate governance framework are embedded in statute.

This is not suggesting that these principles are applied in Nigeria as they would in countries with more sophisticated capital markets. Nevertheless, Nigeria is very keen to attract foreign investments into the country and has put in place a robust system of corporate governance.

9.6.1 The Corporate Governance Mechanism in Nigeria

There are several institutions and individuals charged with the responsibility for ensuring effective accountability of public companies in Nigeria. These include:

9.6.1.1 The Government

The Government plays a key role in corporate governance through the promulgation of various laws affecting the management and control of business enterprises in Nigeria, irrespective of their size (Okike, 2007).

The main legal framework for corporate governance is the Companies and Allied Matters Act (CAMA) 2004. Here are some key aspects of the Act as it relates to corporate governance:

Part VIII - Directors and Secretaries of the company

The board of directors provides oversight function over an organization. Section 246 (1) of CAMA requires that all registered companies have at least two directors. The directors are reappointed at the annual general meeting of the company (section 248). Members of the company can by an ordinary resolution remove a director before the expiration of his period of office. Whereas the Act (section 267(1)) states that the remuneration of the directors shall be determined by the company in annual general meeting, in practice, this task is now assigned to remuneration committees, as specified in the Code of Corporate Governance.

Duties of Directors

The major responsibility of directors is to ensure that the organization is managed in the best interest of the stakeholders. They provide entrepreneurial insight and ethical leadership to the organization and must ensure that they have strategies in place for the efficient management of the human, material and financial resources entrusted to them.

Section 279(1) of CAMA states that '*a director of a company stands in a fiduciary directors relationship towards the company and shall observe the utmost good faith towards the company in any transaction with it or on its behalf*''. In section 279(3), a director is required to '*act at all times' in what he believes to be the best interests of the company as a whole so as to preserve its assets, further its business, and promote the purposes for which it is formed, and in such manner as a faithful, diligent, careful and ordinary skilful director would act in the circumstances'*.

Financial Statements and Audit

Directors are accountable to the shareholders who appoint them. The annual report is the means by which directors make themselves accountable to the shareholders. Section 331 (1) of CAMA states 'every company shall cause accounting records to be kept …' and 'the accounting records shall be sufficient to show and explain the transactions of the company' (section 331(2). Section 334 places the responsibility for the preparation of the company's financial statements on its directors. They have a fiduciary duty to prepare financial statements, which represent a true and fair view of the company's financial operations. The shareholders must approve these financial statements not less than 21 days before the Annual General Meeting (AGM) of the company. The directors are also expected to prepare a director's report, which contains information about the entire activities of the company as well as significant changes that may have taken place within the year.

9.6.1.2 The Financial Reporting Council (FRC)

Given the renewed emphasis on effective corporate governance and accountability in Nigeria, and the need to have effective mechanisms for developing and monitoring corporate governance codes and guidelines, the Federal Government passed the Financial Reporting Act, No.6, 2011. The Act established the Financial Reporting Council (FRC) as a government parastatal under the supervision of the Federal Ministry of Industry, Trade and Investment. In addition to developing and publishing accounting and financial reporting standards to be observed in the preparation of financial statements by public listed companies, the FRC is also responsible for developing and enforcing compliance with codes of corporate governance.

In section 11 of the FRC Act, the Council's main objects are to:

a) Protect investors and other stakeholders' interest;
b) Give guidance on issues relating to financial reporting and corporate governance to professional, institutional and regulatory bodies in Nigeria
c) Ensure good corporate governance practices in the public and private sectors of the Nigerian economy
d) Ensure accuracy and reliability of financial reports and corporate disclosures, pursuant to the various laws and regulations currently in existence;

e) Harmonize activities of relevant professional and regulatory bodies as relating to corporate governance and financial reporting.

Additional objects include

f) Promoting the highest standards among auditors and other professionals engaged in the financial reporting process.
g) Enhancing the credibility of financial reporting; and
h) Improving the quality of accountancy and audit services, actuarial, valuation and corporate governance standards.

The FRC operates through the following operating arms:

- Directorate of Accounting Standards – Private Sector
- Directorate of Accounting Standards – Public Sector
- Directorate of Auditing Practice Standards
- Directorate of Actuarial Standards
- Directorate of Valuation Standards
- Directorate of Inspection and Monitoring
- Directorate of Corporate Governance

Through these various arms, the FRC ensures high quality financial reporting and effective monitoring of companies, external auditors and other professionals whose work bear upon financial reporting integrity and corporate governance of entities.

Since the establishment of the FRC, it has issued and revised codes of corporate governance for public listed companies (examined later), the latest being the exposure draft on the Nigerian Code of Corporate Governance 2018. The inspection and monitoring unit has also been actively monitoring compliance with accounting and auditing standards, including compliance with the code of corporate governance.

9.6.1.3 The Corporate Affairs Commission (CAC)

The Corporate Affairs Commission (CAC) was established as an autonomous body by the Companies and Allied Matters Act (CAMA) 1990 (section 1) to oversee the regulation and supervision of the formation, incorporation, registration, management and winding up of companies. The CAC replaced the Companies Registry because it was found to be grossly deficient in discharging its responsibilities. It was a department of the Federal Ministry of Commerce and Tourism, responsible for the registration and administration of the repealed Companies Act 1968.

All companies are required to submit their audited financial statements to the CAC within 42 days of the annual general meeting. Small companies may deliver modified statements and balance sheets to the CAC. The Registrar of Companies at the CAC monitors compliance with the requirements of the Act and specifies penalties, albeit dated, for non-compliance by companies and their officers. In section 7(c), the CAC is also charged with the responsibility to 'arrange or conduct an investigation into the affairs of any company where the interests of the shareholders and the public so demand'. The Commission can provide information about any company on request, through its Wide Area Network System. The availability of information to the public increases confidence in the business community and provides protection for the individual consumer, creditor or shareholder (Okike, 2007).

Whereas Wallace (1987); Okike (1995, 1999); ROSC, (2004) all criticized the role of the CAC as being perfunctory and ineffective, there has been a lot of rebranding exercise within the CAC, to make it more investor friendly, and reduce the cost of doing business in Nigeria.

Part of this rebranding is the Companies Regulation 2012, approved by the Minister of Trade and Investment pursuant to Sections 16, 585 and 609 of the CAMA 1990. The Regulation came into effect from 1 January 2013.

9.6.1.4 The Securities and Exchange Commission (SEC)

The Securities and Exchange Commission (SEC) is a Government Agency responsible for the regulation and development of the Nigerian Capital Market. It replaced the Capital Issues Commission, and was established by Decree No.71 of 1979, and effective retrospectively from 1 April 1978. The SEC plays an important role in corporate governance in relation to companies listed on the capital market (Okike, 2007). A series of reviews of the country's capital market and the financial system led to the promulgation of the Investments and Securities Decree No.45 of 1999 (later referred to as the investment and Securities Act of 1999). This Act was promulgated to make the investment climate in Nigeria more appealing to foreign investors (Okike, 2007). However, the Investments and Securities Act 2007 has repealed the Investments and Securities Act 1999. The Investment and Securities Act 2007 came into force on 25 June 2007, and provides for the establishment of the Securities and Exchange Commission as the apex regulatory authority for the Nigerian Capital Market. It empowers the SEC to regulate the market to ensure the protection of investors, maintain fair, efficient and transparent market and reduce systemic risk. The Act brings with it, a set of new market infrastructures and wide-ranging system of regulation on investment and securities business in Nigeria, especially in the area of mergers, acquisitions and take-overs, and collective investment Schemes, where new provisions have been made.

The SEC was primarily established to protect the interest of investors against fraudulent and unwholesome practices of stockbrokers and other intermediaries (Orji, 2000). It regulates securities market participants under the Investment and Securities Act of 2007, and the Securities and Exchange Commission Rules and Regulations 1999.

Amongst other things, the roles of the Commission include keeping and maintaining separate registers of foreign direct investment and portfolio investments in Nigeria, promoting the education of investors and the training of intermediaries in the securities industry.

As a member of the International Organisation of Securities Commissions (IOSCO), the SEC is keen to promote high standards of regulation to maintain a sound and efficient securities market in Nigeria. As a result it has an Enforcement and Compliance Department, which is responsible for administering the Commission's enforcement programme. There is evidence of enforcement on the Commission's website and in some cases defaulters are brought before the Administrative Proceedings Committee (APC), a quasi-judicial court, with only civil jurisdiction. Appeals against decisions of the APC are made at the Investment and Securities Tribunal (IST). Enforcement action can take many forms, including payment of fines, a ban, suspension, or even forwarding the case to the Nigerian Police, the Economic and Financial Crimes Commission (EFCC) or the Attorney-General of the Federation, where criminal activity is involved.

9.6.1.5 The Nigerian Stock Exchange (NSE)

The Nigerian Stock Exchange (NSE) is a registered company limited by guarantee and licensed under the Investments and Securities Acts 2007. Originally established by the Nigerian Stock Exchange Act of 1961, the NSE is self-regulating and plays a crucial role in the mobilisation of capital. The rules of the Stock Exchange contain conditions for trading and admission of securities on the floor of the Exchange.

Although Nigeria has an established Stock Exchange, it cannot be described as a Stock exchange based financial system (Demirag, 1998) because it does not play a significant role in the mobilisation of savings. Only a few companies use it for new issues. For instance, there are currently under 300 companies listed on the NSE, even though there are over 600,000 companies registered in Nigeria. In a typical Stock Exchange based financial system, the stock market is a major source of equity and other finance, and provides a market for corporate control (Okike, 2007). However, the management of the NSE aspires to make it the ultimate Stock Exchange for championing the development of Africa's financial markets[95].

The NSE continuously encourages and supports listed companies to establish global best practices with respect to corporate governance for the benefit of the listed companies, the continuous development of the Nigerian capital market and the sustainable development of the economy. Also, the NSE is committed to adopting the highest levels of international standards and belongs to a number of international and regional organizations that promote the development and integration of global best practices across its operations. The NSE is a member of the international Organization of Securities Commissions (IOSCO), the World Federation of Exchanges (WFE), Sustainable Stock Exchanges (SSE) Initiative, the SIIA's Financial Information Services Division (FISD) and the Inter-market Surveillance Group (ISG). It is also a founding member of the African Securities Exchanges Association (ASEA).

With the establishment of the FRC, some of the shortcomings of the NSE in relation to monitoring corporate disclosures (Okike, 2007) are now under control.

9.6.1.6 The Society for Corporate Governance (SCG)

The Society for Corporate Governance in Nigeria is a not for profit organization registered to foster the implementation of corporate governance standards in Nigeria and other emerging economies. The organization equally aims at promoting corporate governance practices locally and internationally. They are saddled with the responsibility of organizing seminars, trainings and workshops with the aim of providing education on matters relating to corporate governance. They are involved in research on corporate governance, which has culminated in certain publications such as the Bi-annual Journal of Corporate Governance. The Society for Corporate Governance organizes the following programmes:

- Being an Effective Member of an Audit Committee
- Strategic Corporate Social Responsibility
- Company Secretaries and Board Effectiveness
- Board Enhancement Programmes
- Board Evaluation
- Leading an Effective Board
- Annual Conference on Corporate Governance
- Board Strategy Session

9.6.1.7 The Institute for Corporate Governance in Nigeria (ICGN)

The Institute of Corporate Governance in Nigeria (ICGN) is a professional body formed by a team of experts from different fields with the aim of promoting best corporate governance practices and business ethics standards in Nigeria, Africa and globally. The Institute was established under the Companies and Allied Matters Act CAP 59 of 1990 of the Federal Republic of Nigeria, as approved by the Federal Ministry of Justice, Federal Ministry of Education and the Corporate Affairs Commission. They embark on research, organize trainings, seminars, workshops on corporate governance and provide certifications in corporate governance. There are two classes of certifications awarded by the institute namely, the Certified Corporate Governance Professional and the Governance Risk Management and Compliance Professional. The Institute equally publishes a journal titled: *Corporate Standards: An International Journal of Corporate Governance Research*, discussion papers and newsletters.

9.6.1.8 Auditors

Auditors play an important role in corporate governance irrespective of the system of corporate governance in place. Their role in lending credibility to management prepared financial statements is often a legal requirement. In Nigeria, section 359(1) of the CAMA 1990 requires that auditors "make a report ... on the accounts examined by them..." and express their opinion on 1) whether the company whose accounts are to be examined has kept proper books of account. 2) whether the accounts being examined by the auditor give a true and fair view of the state of affairs of the company; and 3) whether the company has complied with statutory and other disclosure requirements as stipulated by CAMA 1990. However, the overriding statutory requirement is that the profit and loss account and balance sheet give a true and fair view of the results and state of affairs of the company (Okike, 2007). Therefore, through the independent examination of the books and records of the company, the audit function is part of the mechanism for enhancing confidence in corporate annual reports.

Given the onerous professional, legal and social responsibility of the role of statutory auditors, they are required to have appropriate training and experience, and are also ethically bound to conduct their audit in accordance with recognized procedures and standards (Okike, 2007). Prior to the promulgation of the CAMA 1990, the Institute of Chartered Accountants of Nigeria (ICAN) enjoyed the monopoly of being the only professional accountancy body recognized to determine the standards and skill necessary to be attained by those seeking to practice as auditors. However, following the enactment of CAMA 1990, the monopoly enjoyed by ICAN was removed, through the provision of section 358(1) of CAMA 1990, which states "a person shall not be qualified for appointment as an auditor of a company for the purposes of this Act, unless he is a member of a body of accountants in Nigeria, established from time to time by an Act or Decree". This provision brought about some discontent within the accounting profession, as members of ICAN held the view that members of the Association of National Accountants (ANAN) lacked the technical competence to perform audits[96]. This enactment gives statutory recognition to ANAN, which in effect means there are now two professional accounting bodies in Nigeria (Okike, 2007), both of which have two representatives each on the Financial Reporting Council. Through their membership of the FRC, ICAN and ANAN can collaboratively promote high standards of accountability and corporate governance in Nigeria.

9.6.1.9 Audit Committees

Whilst it is true that as a former colony of Great Britain, Nigeria imbibed the laws and corporate practices of the UK, a revision of the Companies Act in 1990, provided the opportunity for Nigeria to incorporate into law, corporate governance practices that were suited to the country's needs. One of such peculiar initiatives was the requirement of Section 359(3) of CAMA 1990 that auditors of public companies make a report to an audit committee, which shall be established by all public companies. The audit committee shall examine the auditors' report and make recommendations thereon to the annual general meeting as it may think fit. This requirement was made before the Cadbury Committee was established in 1991. This influence came rather from the United States where the SEC had made a requirement that from 1988 all SEC regulated companies should have an audit committee as part of the listing requirement in response to the Treadway Report on Fraudulent financial reporting in 1987 (Okike, 2007).

There has been a growing interest in the use of audit committees in recent times, and it is accepted as good business practice in many countries, with the key objective of raising the standard of corporate governance. Effective audit committees provide assurance to shareholders that the auditors, who act on their behalf, are in a position to do so and do safeguard their interests. In the UK, the Cadbury Report 1992 recommended that all listed companies should establish audit committees. But before this recommendation, there was evidence that during 1991, 53 per cent of top 250 listed companies in the UK had effective audit committees. But following the recommendation of the Cadbury Committee the percentage had risen to 85 per cent of the larger and 83 per cent of the smaller listed companies (Collier, 1992, 1996).

CAMA 1990 became effective on 2 January 1990, and companies were required to comply with its provision by this time. But Okike (2002) provides evidence that many listed companies had still not complied in 1993, three years after the promulgation of the Act. This was as a result of the weaknesses identified in the corporate monitoring mechanisms in Nigeria (Okike, 2007). It is unlikely that such weaknesses will occur in the new corporate governance era, with the establishment of the FRC, which has ensuring good corporate governance practices in Nigeria, as one of its objectives.

Section 359(4) of CAMA 1990 stipulates that the audit committee should consist of an equal number of directors and representatives of the shareholders of the company (subject to a maximum of six members). By the provision of this section, the law gave a formal recognition to the body of shareholders in Nigeria (Okike, 2007). Para.6.4 of the draft Code of Corporate Governance reiterates the provisions of CAMA 1990 about having an audit committee, all of whose members should be financially literate. Majority of the members of the audit committee are expected to be independent non-executive directors. The Chairman should be elected amongst its members, and the committee must meet at least once every quarter.

In section 359(6)) of CAMA 1990 and Para.6.4.6 of the Code of Corporate Governance 2018, the responsibilities of the statutory audit committee are to:

(a) ascertain whether the accounting and reporting policies of the company are in accordance with legal requirements and agreed ethical practices;
(b) review the scope and planning of audit requirements;

(c) review the findings on management matters in conjunction with the external auditor and departmental responses thereon;
(d) keep under review the effectiveness of the company's system of accounting and internal control;
(e) make recommendations to the Board in regard to the appointment, removal and remuneration of the external auditors of the company; and
(f) authorize the internal auditor to carry out investigations into any activities of the company which may be of interest or concern to the committee.

The Code of Corporate Governance 2018 (Para.**6.4.7**) imposes additional responsibilities on the audit committee, which are to:

> 6.4.7.1 Exercise oversight over management's proceedings to ascertain the integrity of the Company's financial statements, compliance with all applicable legal and other regulatory requirements, and assess the qualifications and independence of the external auditors, the performance of the Company's internal audit function as well as that of the external auditors.
>
> 6.4.7.2 Ensure the establishment of and exercise oversight on the internal audit function which provides assurance on the effectiveness of the internal controls…
>
> 6.4.7.3 Ensure the development of a comprehensive internal control framework for the Company, obtain appropriate (internal and/or external) assurance and report annually in the Company's audited financial report, on the design and operating effectiveness of the Company's internal controls over the financial reporting systems.
>
> 6.4.7.4 Oversee the process for the identification of fraud risks across the Company and ensure that adequate

prevention, detection and reporting mechanisms are in place.

6.4.7.5 Discuss the interim or annual audited financial statements as well as significant financial reporting findings and recommendations with management and external auditors prior to recommending same to the Board for their consideration and appropriate action.

6.4.7.6 Maintain oversight of financial and non-financial reporting

6.4.7.7 Review and ensure that adequate whistle-blowing policies and procedures are in place and that the summary of issues reported through the whistle-blowing mechanism are highlighted to the committee.

6.4.7.8 Review, with the external auditors, any audit scope limitations or significant matters encountered and management's responses to same.

6.4.7.9 Develop a policy on the nature, extent and terms under which the external auditors may perform non-audit services.

6.4.7.10 Review the independence of the external auditors in line with the policy referred to in 6.4.7.8 above prior to their appointment to perform non-audit services to ensure that where approved non-audit services are provided by the external auditors, there is no real or perceived conflict of interest, or other legal or ethical impediment.

6.4.7.11 Preserve auditor independence, by setting clear hiring policies for employees or former employees of external auditors

6.4.7.12 Ensure the development of a Related Party Transactions policy and monitor its implementation by management. The Committee should consider any related party transaction that may arise within the Company.

6.4.8 The committee should, at least once in a year, include on its agenda, a discussion with the head of the internal audit function and the external auditors without management being present to facilitate an exchange of views and concerns that may not be appropriate for open discussion.

Whilst the establishment of audit committees may not be the panacea for all corporate malpractices, nevertheless, it is expected that the presence of such a committee would encourage improved audit quality, enhance auditor's independence, strengthen the financial reporting process and lessen the potential for fraudulent financial reporting (Okike, 2007). The anti-corruption initiatives of the Federal Government and the support of other bodies like the NSE, the SEC, ICAN and ANAN, are likely to lend impetus to the effectiveness of audit committees.

9.6.1.10 Shareholders' Associations

The indigenisation exercise of 1972 (see Okike, 1995) saw the shares of many foreign nationals change hands and many Nigerians bought shares from companies listed on the NSE. Thus the number of indigenous shareholders increased on the stock market. This rise in shareholder numbers also gave rise to activists shareholders, who attended annual general meetings to cause trouble. This shareholder group often challenged corporate boards whenever they were dissatisfied with the performance of the companies in which they held shares. They soon formed themselves into formidable groups.

As mentioned earlier, section 359(4) CAMA 1990 stipulates that the audit committee shall consist of an equal number of directors and representatives of the shareholders of the company (subject to a maximum of six members). Besides giving formal recognition to shareholder associations in Nigeria, the provision of this section of the Act led to the proliferation of shareholder associations and notoriety of shareholder activism in Nigeria. As Adegbite (2010) reports,

> 'there is no doubt that the setting up of shareholder associations was encouraged due to the need to coordinate several small, passive and dispersed shareholders; however the intended activism has been hijacked by individuals whose aims are to reap personal benefits, which is truly characteristic of the broader political culture of the country. In the quest of achieving this, several senior executives of shareholders associations bully corporate management through threats of AGM disruptions and negative media propaganda and have thus constituted themselves "terrorist" gangs" who are now feared by corporate executives... Given that shareholders' associations have become constitutionally empowered to challenge managerial and board excesses, they constitute a great threat to the status quo of traditionally unchecked corporate corruption and governance malfunction, if their powers are applied positively'.

Unfortunately, the reality is that rather than shareholder activism playing the role of curbing management misdemeanours and protecting the interests of shareholders, some of the executives of these shareholder associations have become too familiar with the corporate boards of the companies in which they hold shares. The consequence is that they have become ineffective in their activist role, which alternatively means that corporate governance is not as effective as it could be.

In order to address the observed ineffectiveness of the shareholders' association and the negative impact of their activities (Adegbite, 2010), the SEC introduced a Code of Conduct for Shareholders' Associations in Nigeria. According to a report in the Sun News (2007) (quoted in Adegbite, 2010), some of the identified key problems necessitating the need for such a Code was the '*proliferation of shareholder associations, concern over behaviour of some members at Annual General Meetings, intense competition towards getting on companies' audit committees, governance problems and unclear succession arrangements and the inadequate members enlightenment on shareholders rights, privileges and responsibilities...'*

Given that the quest to be on the audit committees of their investee companies is one of the reasons for the proliferation of shareholder associations, one important recommendation of the Code is that executive members of the shareholder associations cannot be elected as audit committee members. The establishment of this Code of Conduct is another example of Nigeria adopting codes and passing laws that are suited to the country's socio-political and cultural challenges. It is described as '*a code ... designed to ensure that association members uphold high ethical standards and make positive contributions in ensuring that the affairs of public companies are run in an ethical and transparent manner and also in compliance with the Code of Corporate Governance for public companies*'. Whilst the UK has a Code directed at shareholders, in this case, institutional investors, the emphasis is different. The UK Stewardship Code was introduced in 2010 (and has been revised in 2012 and more recently in 2018), to "enhance the quality of engagement between investors and companies to help improve long-term risk-adjusted returns to shareholders". It is directed at asset managers who hold voting rights on shares in UK companies. The aim is to make institutional investors, who manage the assets of others to engage in corporate governance in the interests of their beneficiaries.

9.7 Development of Corporate Governance Codes in Nigeria

Given the spate of corporate scandals especially in the UK (BCCI, Polly Peck, Maxwell's Mirror Group) in the 1990s, the need to ensure effective governance of corporations received fresh impetus within the global business community, including Nigeria. Having been influenced by the development of corporate governance codes in other countries, especially the Cadbury Report of 1992 and King Report 1994, the Securities and Exchange Commission (SEC) set up The Committee on Corporate Governance of Public Companies in Nigeria

> To review the practices of corporate governance in Nigeria and thereafter, recommend a Code of Best Practices to be followed by public companies registered in Nigeria in the exercise of power over the direction of the enterprise, the supervision of executive actions, the transparency and accountability in governance of these companies within the regulatory framework and market.

The Committee made up of 17 members[97], headed by Mr Atedo Peterside, submitted its Report in April 2001, in which it made recommendations, after a thorough review of existing practices in Nigeria and other countries. Their recommendations reflected global best practices and focused on transparency and accountability of the management and boards of public companies (Okike, 2007). Acting on their recommendations, the SEC issued the Code of Best Practices on Corporate Governance in Nigeria (SEC Code) in October 2003.

9.7.1 Securities and Exchange Commission Code October 2003

This was the first corporate governance Code issued in Nigeria, and it was applicable to all publicly quoted companies. The Code emphasised the role of the board of directors and management, shareholder rights and privileges and the audit committee. The Code greatly impacted corporate governance in Nigeria, but was not quick in responding to the changing economic climate that demanded that it be reviewed and updated. The stagnation and inadequacy of the 2003 SEC Code prompted regulators in specific industry sectors to issue industry-specific corporate governance codes, which responded to the immediate demands of the changing business environment in their specific sectors.

9.7.2 Central Bank of Nigeria Code of Corporate Governance for Banks in Nigeria Post-Consolidation 2006

The Central Bank of Nigeria (CBN) was the first industry specific regulator to address the deficiencies of the SEC Code. In 2006, the CBN issued the Code of Corporate Governance for Banks in Nigeria Post-Consolidation (CBN Code). Compliance with the provisions of this Code was mandatory for all banks operating in Nigeria. The consolidation of banks in Nigeria in 2005 made the issuance of the CBN Code essential because there was the need to address the identified weaknesses in the corporate governance of banks and the governance challenges that would arise post-consolidation. Also, the SEC 2003 Code did not address some of the industry specific problems peculiar to the financial sector. The CBN Code was issued to ensure that bank Chief Executive Officers were more accountable. It carried penalties for non-compliance, including jail terms for erring CEOs. It also emphasized effective risk management measures and highlighted the importance of the role of internal auditors. Other industry specific codes followed suit.

9.7.3 National Pension Commission (PENCOM) Code of Corporate Governance Practices for the Insurance Industry in Nigeria 2008

In 2008, the National Pension Commission (PENCOM) issued the Code of Corporate Governance for Licensed Pension Operators to guide pension fund administrators (PFAs) as well as pension fund custodians (PFCs) on best corporate governance practices. The Code establishes rules based on best practices to guide PFAs and PFCs on the structures and processes required to achieve optimal governance set up. It is developed with a view to establishing overall economic performance and market integrity as it creates incentives for the pension scheme to impact positively on the stakeholders, who are directly affected by the pension reform, and gain their confidence.

Also, the Code promotes transparent and efficient implementation of the scheme by all the operators. The aim is to encourage self-regulation by providing a common value system among the operators. The Code is based on internationally accepted principles of good corporate governance and its requirements are consistent with the provisions of the Pension Reform Act 2004, rules, regulations and guidelines issued by the Commission and are also considered transparent and enforceable.

9.7.4 The Code of Good Corporate Governance for the Insurance Industry in Nigeria 2009

The role of the insurance industry in the protection of the national economy is undeniable, hence the need for an effective and efficient regulatory system to supervise the activities of the insurance industry. Such supervision was introduced in Nigeria on a modest scale under the provisions of the Insurance Act 1961[98]. As the industry grew and became more sophisticated, a more robust regulatory apparatus was put in place in the form of National Insurance Commission (NAICOM).

As part of its efforts to ensure the industry was keeping abreast of developments within the global economy, and was being operated efficiently and effectively, in 2009 NAICOM issued the Code of Good Corporate Governance for the Insurance Industry in Nigeria. NAICOM believes that a good code of corporate governance will be a major recipe for unleashing the hidden potential of the industry for maximum impact on the national economy.

9.7.5 The Code of Corporate Governance in Nigeria 2011

The 2003 SEC Code did not contain adequate provisions on certain contemporary corporate governance issues relating to independent directors, critical board committees, directors' appointment, tenure, remuneration and evaluation, the independence of the external auditors, whistle-blowing procedures, sustainability issues; and general disclosure and transparency issues (Ofo, 2013). Appropriately, the SEC 2003 Code was updated and the Securities and Exchange Commission issued a new Code of Corporate Governance in Nigeria on 1 April 2011. The Code addressed five main principles, including leadership, effectiveness of board structures, accountability, remuneration and relations with shareholders. Shortly after the 2011 SEC Code of Corporate Governance became functional, the Financial Reporting Council (FRC) of Nigeria Act was promulgated.

9.7.6 The Code of Corporate Governance for the Telecommunication Industry 2016

In response to the emerging corporate governance challenges within their specific sectors, a number of other corporate governance codes were issued. These include the Code of Corporate Governance for the Telecommunication Industry 2016, issued by the Nigerian Communications Commission (NCC). This replaced the NCC Code of 2014.

Also, the CBN Code of Corporate Governance for Banks and Discount Houses in Nigeria 2014 has since replaced the CBN Code of 2006.

The Financial Reporting Council was saddled with the responsibility of setting up a Directorate of Corporate Governance to handle all corporate governance issues in the country. In an effort to address the multiplicity of overlapping legislations, this directorate made an attempt to unify the industry sector codes and developed the National Code of Corporate Governance (NCCG), which was to take effect from 17 October 2016. This national code was more robust than the previous ones because it provided a code not only for public entities but also private entities as well as not-for-profit organizations. However, following stiff opposition from various stakeholders, the Federal Government suspended this National Code of Corporate Governance on 28 October 2016. It appears that the socio-cultural and political issues examined in Okike (1994, 2004) surfaced again, this time, also crossing religious boundaries hence the suspension of the NCCG. The situation was even more precarious because not only was the NCCG suspended, the Minister of Industry, Trade and Investment, Okechukwu Enelamah, issued a query to the Financial Reporting Council Nigeria to explain the reason d'entre of the Code.

Given that the remit of the FRC includes developing the mechanism for the periodic assessment of the codes and guidelines issued by the Council, on 18 January 2018, a fifteen-man committee formed by the board of the FRC, proposed the Nigerian Code of Corporate Governance (NCCG) 2018. Companies are required to commence financial reporting and applying the new code in their financial reports starting in the financial year ending on or before January 1, 2020. However, earlier adoption is encouraged. This new Code embraces all public companies, private companies that are holding companies of public companies, concessioned and/or privatised companies as well as regulated private companies who file returns to any other regulatory authority other than the Federal Inland Revenue Services and the Corporate Affairs Commission. Compliance will be monitored by the FRC through the NSE, and other regulatory bodies and associations.

The portion of the code relating to not-for-profit organizations remains suspended.

9.7.7 Nigerian Code of Corporate Governance (NCCG) 2018

The FRC describes NCCG 2018 as one that seeks to institutionalize the highest standards of corporate governance best practices in Nigerian companies, especially those companies without industry sector codes or regulations. The Code is also meant to promote public awareness of essential corporate values and ethical practices that will enhance the integrity of the market and rebuild public trust and confidence in the Nigerian economy.

Given that the Code is directed at companies of various sizes and within diverse industrial sectors, the need for flexibility and scalability was taken into consideration in determining that a 'principles-based' approach is preferable to a 'rules-based' approach. Furthermore, the implementation of the Code is based on 'Apply and Explain' principle, rather than the 'comply and explain' approach adopted in the UK. Companies are required to apply the principles laid down in the Code and explain how they have applied them. The decision to adopt this approach was influenced by practices in other countries, which require voluntary adoption of codes of corporate governance, especially Mauritius and South Africa, two other Commonwealth countries, which have also adopted the 'Apply and Explain' principle.

The Code philosophy is explained as follows:

Apply principles: These principles are those fundamental to good governance, which all companies should strive to adopt.

Explain Principles: Companies are required to provide explanations for those recommended practices that have been applied, or explanations for how these or any other have yielded expected outcomes.

The NCCG consists of 7 parts and 28 principles, detailing the recommended practical application of the Code. These are outlined below:

A) *Board of Directors*. This part contains 16 principles of what is expected of the Board of Directors of a company, its composition and structure, and the roles and responsibilities of executive and non-executive members of the Board and the Company Secretary. It also details procedures for ensuring the smooth and effective running of the Board, including induction and training programmes for members of the Board.

B) *Assurance*. This part consists of four principles, which draw attention to the importance of having a sound framework for the management of risk and effective internal control systems, including the role of internal and external auditors, in providing assurance as to the reliability of management's prepared financial statements.

C) *Relationship with shareholders*. The three principles contained in this part stress the importance of good governance through the maintenance of a good relationship with shareholders. The annual meetings provide the shareholders with the opportunity to gain a deeper understanding of the company, its governance and performance, and also enables them exercise their rights as owners. Regular dialogues with shareholders help balance their needs with company objectives. All shareholders should be treated equally and the interests of minority shareholders need to be protected.

D) *Business Conduct with Ethics*. This part of the Code deals with principles to ensure that companies carry out their affairs in the most ethical manner, not only to enhance and protect their reputation, but also to boost the confidence of investors. It also requires companies to have procedures in place for dealing with any conflicts of interest, corrupt practices and insider trading.

E) *Sustainability*. This important principle emphasizes the need for companies to be socially responsible, and to be good corporate citizens, contributing to economic development.

F) *Transparency.* When a company is in constant communication with key stakeholders and transparent about its activities, it enables them make informed decisions. So also does full and comprehensive disclosure of material matters to investors and stakeholders promote good corporate governance. The Code requires companies to be transparent about their affairs to shareholders and key stakeholders.

Companies must aim to create value for their shareholders by putting in place appropriate strategies and policies that would enable them achieve their objectives. They must commit to effective management of all human, material and financial resources, and have effective boards in place, consisting of people with integrity and high technical competence. They must engage with their shareholders and maintain dialogue with key stakeholders. Having these appropriate structures in place will signal to the market that these companies are well managed, which could be attractive to potential investors.

Although NCCG 2018 is voluntary, adherence to the principles articulated in the Code, including commitment to maintaining good governance practices increases levels of transparency, trust and integrity, creates an environment that manages risks effectively, minimizes losses, and makes a business operation sustainable.

9.8 Summary and Conclusion

This chapter is a synopsis of the current state of corporate governance in Nigeria, Africa's largest economy and the third largest member of the Commonwealth of Nations. There is no doubt that given Nigeria's wealth of physical, human and material resources the country has the potential to attract a huge amount of foreign investments into the country. Yet it is a country with many paradoxes. On the one hand it is presumed to be economically significant because of its rich oil wells. Yet at the same time it is classed amongst the countries where poverty is at its height. This is because of endemic corruption in all facets of public and private life. This notwithstanding, there are concerted efforts being made within the corporate sector to make Nigeria the place where serious investors would give first consideration.

The passing of the Companies and Allied Matters Act (CAMA) 2004 (Repeal and Re-enactment) Bill, 2018 and the establishment of the Financial Reporting Council (FRC) are important steps in the right direction. The new CAMA 2018 will promote the ease of doing business in Nigeria, and ensure that corporate developments in the country are responding to developments in the global market. Also, one of the objectives of the FRC is to ensure high quality financial reporting and effective monitoring of companies. Investors would be willing to invest in an economy with appropriate structures in place that promote high standards of accountability and corporate governance. Although as a former colony of Great Britain Nigeria's corporate sector and legal system bore great semblance with that of the UK, there is evidence from the new Bill and the exposure draft of Nigeria Code of Corporate Governance 2018 that there are concerted efforts to ensure that Nigeria's legal system and the regulation of the corporate sector reflect the peculiar socio-political and cultural demands of the country and meet the aspirations of the people. Having established the FRC and given it the mandate to ensure the integrity of the corporate sector, the Federal Government would do well not to interfere with their work. The suspension of the NCCG 2016 is a case in point.

REFERENCES

Adegbite, E. (2010) *The Determinants of Good Corporate Governance: The case of Nigeria,* being a Thesis submitted in fulfilment of the requirements for the Degree of PhD in Management, Faculty of Management, Cass Business School, London.

Anibaba, M.O. (1990), *A Short History of the Accountancy Profession in Nigeria.* West African Publishers Ltd, Lagos

Barbour, K.M., Oguntoyinbo, J.S., Onyemelukwe, J.O and Nwafor, J.C. (1982) *Nigeria in Maps*, Hodder & Stoughton, London.

Collier, P. A. (1992), *Audit Committees in Large UK Companies.* London: Research Board of the Institute of Chartered Accountants in England and Wales.

Collier, P.A (1996) The Rise of the Audit Committee in UK Quoted Companies: A Curious Phenomenon, *Accounting, Business and Financial History,* 6: 121-140.

Demirag, I.S. (1998) "Short Termism, Financial Systems, and Corporate Governance: A Theoretical Framework". In Demirag, I.S (ed.) *Corporate Governance, Accountability, and Pressures for Reform: An International Study.* JAI press Inc.

Financial Reporting Council of Nigeria (FRCN) (2018) *The Nigerian Code of Corporate Governance 2018*, Abuja, Nigeria.

Franks, J. and Meyer, C. (1994) Corporate Control: A Comparison of Insider and Outside Systems, Working Paper, London Business School.

Irukwu, J.O (2009) in the Foreword to the Code https://www2.deloitte.com/content/dam/Deloitte/ng/Documents/centre-for-corporate-governance/code-of-good-corporate-governance-for-the-insurance-industry-in-nigeria.pdf

King Report (1994) *The King Report on Corporate Governance*, Institute of Directors in South Africa, South Africa, 29 November.

Ofo, N. (2013, January 7). *Historical Development of Corporate Governance in Nigeria.* Retrieved from The Corporate Prof. Website: http://thecorporateprof.com/historical-development-of-corporate-governance-in-nigeria/

Okike, E.N.M. (1995), "Corporate Audit Report and the Structure of the Market for Audit Services to Listed Companies in Nigeria: A Longitudinal Study". A Research Report submitted to the Research Committee of the Institute of Chartered Accountants of Nigeria (ICAN).

Okike, E.N.M. (1998), "Voluntary Disclosures in Corporate Audit Reports: A Nigerian Case Study (1978-1989)" *The International Journal of Auditing*, 2: 139-166.

Okike, E.N.M. (1999), "A Longitudinal Analysis of the Quality of Audit Reporting on the Accounts of Listed Companies in Nigeria following the Indigenization of Equity Capital (1978-1989)" *Research in Accounting in Emerging Economies, Supplement 1:*255-286.

Okike, E.N.M. (2002) "Influences on the Corporate Governance Framework and Reporting on Listed Companies in a Developing Economy: A Nigerian Case Study", *Paper Presented at the 3rd International conference for the British Accounting Association Special Interest Group in Corporate Governance, at Queen's University Belfast, 16 December.*

Okike, E.N.M. (2004), "Management of Crisis: The Response of the Auditing Profession in Nigeria to the Challenge to its Legitimacy" *Accounting, Auditing & Accountability Journal*, 17(5): 705-730.

Okike, E.N.M. (2007), "Corporate Governance in Nigeria: The Status Quo", *Accounting, Auditing & Accountability Journal*, 17(5): 705-730.

Okike, E.N.M. and Adegbite, E. (2012), "The Code of Corporate Governance in Nigeria: Efficiency Gains or Social Legitimation?", *Corporate Ownership & Control*, 9(3): 262-275.

Okougbo , P. O., & Okike , E. N. (2015). Corporate governance and earnings management: Empirical evidence from Nigeria. *Corporate Ownership and Control, 12*(4), 312-326. doi:10.22495/cocv12i4c2p7

Olutuyi, O. (2017, July 27). *How corporate governance can affect Nigeria's development.* Retrieved from A. Financial Nigeria Development and Finance:

http://www.financialnigeria.com/how-corporate-governance-can-affect-nigeria-s-development-blog-273.html [accessed 12 August 2018]

Orji, N.A. (2000) "The Legal Environment for the Capital Market", *Paper presented at the Workshop on Legislating for the Capital Market in a Democratic Environment, held at Nicon Hilton Hotel, Abuja, 12 October.*

Parker, R.H. (1958) "Company Law in Nigeria", *Accountant*, 139:693-694.

Report on the Observance of Standards and Codes (ROSC) Nigeria (2004) World Bank,

http://www.worldbank.org/ifa/rosc_aa.html

Wallace, R.S.O. (1987) *Disclosure of Accounting Information in Developing Countries: A Case Study of Nigeria*, Doctoral Thesis, University of Exeter.

Wallace, R.S.O (1989) *Accounting and Financial Reporting in Nigeria: A Study Prepared for the Research Board of the Institute of Chartered Accountants in England and Wales.* London, ICAEW.

Endnotes

[87] www.worldometers.info/world-population/nigeria-population/ [accessed 12 August 2018]

[88] https://www.indexmundi.com/g/r.aspx?v=69[accessed 15 August 2018]

[89] https://www.transparency.org/country/NGA [accessed 12 August 2018]

[90] In addition to the concerns, practical company litigation and procedural matters necessitated the review of the 1968 Companies Act. For example, the paucity of Nigerian cases on Company Law meant that 30 years after independence, Nigeria still referred to England for legal inspiration. A lot of cost was involved in obtaining English law reports and text books from Britain.

[91] Nigeria is the only country in the world that has ever made such a provision.

[92] https://medium.com/.../summary-benefits-of-the-companies-and-allied-matters-act-ca. [accessed 15 August 2018]

[93] https://medium.com/.../summary-benefits-of-the-companies-and-allied-matters-act-ca. [accessed 15 August 2018]

[94] https://medium.com/.../summary-benefits-of-the-companies-and-allied-matters-act-ca. [accessed 15 August 2018]

[95] www.nse.com.ng/about-us/about-the-nse/corporate-overview

[accessed 15 August 2018]

[96] Wallace, 1987 and 1992 provides background information to the establishment of ANAN and the rift between ICAN and ANAN.

[97] Members of the Committee were drawn from the public and private sectors, including representatives from the Nigerian Stock Exchange, the SEC and the Corporate Affairs Commission.

[98] Irukwu, J.O (2009) in the Foreword to the Code https://www2.deloitte.com/content/dam/Deloitte/ng/Documents/centre-

for-corporate-governance/code-of-good-corporate-governance-for-the-insurance-industry-in-nigeria.pdf [accessed 18 August 2018]

Chapter 10

Corporate Governance in Pakistan

Beenish Amer

10.0 Introduction

The British Commonwealth is an inter-governmental organization established in 1965. The main objective of this organization is to promote democracy, human rights, good governance, rule of law, free trade, international relationship and world peace. Pakistan became an Islamic state, after the partitioning of the UK's Indian Empire during its independence in August 1947. Originally, it consisted of two parts: West Pakistan (Pakistan) and East Pakistan (now Bangladesh), separated by 1,600 km of Indian territory. The partition was followed by war with India over Kashmir and the mass migration of Muslims, Hindus and Sikhs to resettle within the new borders – an upheaval which led to violence, financial loss and death on a large scale. With the arrival of Indian Muslims and departure of Pakistan's Hindus and Sikhs, Pakistan became almost an entirely Muslim society. Jinnah, honoured as the Quaid-i-Azam or great leader and Pakistan's first Governor-General, died in 1948.

Pakistan became part of the Commonwealth in 1947, at the time of its independence. After 25 years, in 1972, when Pakistan left this organization, Bangladesh was recognized by other member countries. However, Pakistan returned after the democratic elections of 1989. Following the overthrow of the democratically elected government in October 1999, the country was suspended from the councils of the Commonwealth pending the restoration of democracy.

Pakistan's membership of the Commonwealth was suspended from 1999 to 2004, a grouping centred around Britain and its former colonies, following Musharraf's imposition of the emergency rule and the detention of thousands of opposition activists. Musharraf's leadership has seen his country excluded from the Commonwealth more than once.

10.1 Corporate Governance in Pakistan

Pakistan's economy remains in the doldrums, as it still faces clinical and structural challenges. The economic data (Fiscal and Monetary) clearly indicate increased suffering and social unrest due to the unstable political and economic environment caused by an old-fashioned economic growth strategy. Since independence, the pace of growth has averaged above 4.50 percent. But statistics suggest that Pakistan is only a notch above the poorest countries in the World. It has a high poverty rate, low literacy rate, poor infrastructure and outdated agricultural technology, lack of industrialization, wide trade gap, poor healthcare system and low standard of living. If we take a closer look at the present economic indicators, Pakistan's economy is dangerously imbalanced.

Unfortunately, in spite of over seventy years of independence, the country's economic dependency is solely on old methods of relying on foreign inflows and low yielding agricultural sector growth that frequently demands government assistance in support of increases in the prices of food. It is regrettable for the economy that there is a minor reliance on a partially developed industrialized manufacturing sector that does not make serious contribution in this progressive era, which is now entering the fourth industrial revolution. Strong growth demands attention to firm planning and consistency in policy implementation that always remained low priority areas for the decision makers for unknown reasons.

Developed countries have strong mechanisms in place for monitoring the activities of their corporations, embedded in their Corporate Governance Codes. In comparison, Pakistani companies still have weak internal and external corporate governance mechanisms, a situation which has worsened as a result of corporate financial scandals over the last decade. Large companies like Pakistan Telecommunication Company Limited (PTCL) and Pakistan based multi-national company, ENGRO were found to be involved in big scams. Some corporate giants were making profits at the expense of the investors. The situation calls for transparency and external pressure through stakeholders in the absence of which more such scams are likely to result in decreased overall economic development. There is a need for shielding the corporate sector of Pakistan which is the building block of its economy. It is highly undesirable to ignore examining the causes and effects of such weak governance of companies. The corporate governance standard in this regard could reveal further irregularities of the companies. Additionally, globalization has increased the level of pressure for the country to keep abreast of developments within the global economy. This is an opportunity for a developing country like Pakistan to improve its productivity, enhance markets and develop governance standards. Appropriately an examination of the corporate practices and the monitoring mechanisms of Pakistan's corporate sector is beneficial for overall economic development.

Beenish (2013) states that due to many distortions in Pakistan's economy, market forces do not reward good governance or punish unethical practices. The large part of the undocumented economy discourages promotion of transparency and accountability in the organization. Overall management structure is also not conducive to establishing the norms of good governance. Many companies listed on the Stock Exchange are not fully compliant with the code of good governance. Tightly held ownership, lack of professional skills, missing change agents, audit dependability and overall structural weaknesses are bottlenecks in developing corporate governance in these organizations.

10.2 Causes of poor corporate Governance in Pakistan

It is generally understood that board composition and independence are fundamental issues in corporate Governance. Board independence is an important governance mechanism. But in Pakistan there is no concept of independent non-executive directors (INEDs). Independent directors broadly fit into the overall structure of corporate governance, and are necessary to ensure an effective, balanced board. An independent non- executive director is one who is independent of management. Under the Combined Code, a director should not: i) have been an employee of the group within the last 5 years; ii) have a material business relationship with the company; iii) receive any remuneration from the company other than fees, iv) have a share scheme and performance related pay; v) have close family ties with advisers, directors or senior employees; vi) have significant links with other directors; vii) represent a significant shareholder; viii) have served on the board for more than 9 years. In Pakistan independent non-executive directors are practically non-existent. Here, a lot of companies portray their non-executive directors as independent, but in reality none of them is independent.

Here are some reasons why INEDs in Pakistan are not effective:

Most companies are owned by families or closed groups and they usually try to stay away from directors they are unable to control. Controlling shareholders are keen to run their company in their own way and they dislike any kind of interference externally. Due to the absence of independent non-executive directors in Pakistan listed companies, there is no balance of power on the board because they are not prepared to grant any of their powers to outsiders. It all depends on controlling shareholders, whether they really want to have independent non-executive directors on their company's board or not. Another important cause of ineffective independent non-executive directors in Pakistan is that companies do not pay any compensation or remuneration to them. There is just a fixed fee for them for attending a meeting. Professionals are not keen to serve on the boards of listed companies in order to obtain a small fee for attending a meeting. Infact some of them do not have the luxury of time to attend meetings. To run a company smoothly and successfully, it requires highly skilled and experienced professionals. However, this class of professionals are too busy with their own affairs.

There is a mistaken belief amongst Pakistan's listed companies that if they provide salaries to INEDs their independence will be impaired. Independent non-executive directors are supposed to be professionals like external auditors or institutional investors. But in Pakistan they have no incentive to want serve as INEDs because they perceive no long term benefit in doing so. They are not enthused with the nominal fee they receive. Moreover, INEDs in Pakistan's listed companies do not have the required skills and expertise to contribute efficiently in board proceedings.

10.3 Suggestions to address the deficiencies in the code of Corporate Governance

The Code of Corporate Governance issued by the Securities and Exchange Commission of Pakistan (SECP) is relatively new and needs further development. The SECP developed the first Code of Corporate Governance for Pakistan in March 2002 which is now available on its website after few amendments. It is vital that the Cde be followed not only in letter but in spirit as well. There are few suggestions to address the shortcomings of the Code of Corporate Governance:

- In developed countries INEDs are adequately remunerated, so they have the motivation and commitment to devote time to their organization. However, the main issue is that in Pakistan most of the businesses are family owned; majority of the shares are held by one person, so they do not see the need for an INED. Moreover, having an INED can jeopardise their personal interest. Most often INEDs in Pakistani companies are hired on relationship basis, such as the brother of the main shareholder. Secondly, if INED'S are appointed, their remuneration is very basic, depending on the number of meetings they attend. These are hindrances to the success of corporate governance.

- In developed countries, professional INEDs are hired on the basis of their experience, so they have an understanding of how corporate governance should work. However, in Pakistan, companies are owned by a majority shareholder, and INEDs are selected on relationship basis. This ineffect suggests that their role is perfunctory. Moreover, they lack the competence to act as INEDs. Therefore, lack of adequate training is another issue that mitigates against effective corporate governance in Pakistan.
- The third issue in Pakistani context is that there is no distinction between executive and non-executive directors in relation to the liabilities of the company. There have been instances in Pakistan where wrong decisions were made by management, and all the executives, including INEDs were charged to court. Such incidences further discourage professionals from accepting to serve as INEDs. In Pakistan institutional investors do not take part in voting, this in effect implies that the majority shareholder can appoint anyone to take up the role of INED. Whereas in the USA, the situation is entirely different; not only do institutional investors take active part in voting in the INEDs, but also they fully support them, and after nomination, they also provide professional training on corporate issues.
- Laws and codes are available but are not implemented in the true sense. In Pakistan the majority of directors prefer to be directly responsible for the affairs of their companies and

are not interested in compliance with codes. They do not want to create a culture of independent directors so as to protect their own interest. It is generally acknowledged that implementation of the Code has resulted in greater financial discipline, uniform presentation of financial results and an organized and more disciplined corporate culture which has bridged the gap between various stakeholders, the board of directors and the auditors. Therefore, proper implementation of the code of corporate governance is important for effective governance. It is the responsibility of the board to make sure of its practical implementation. Another important issue is the lack of appropriate mechanisms for monitoring compliance with the codes.

- It is suggested that an Institute of Directors needs to be established, to create a data bank of persons qualified and experienced enough to serve as directors. Several organizations, such as the SECP, the Institute of Chartered Accountants of Pakistan (ICAP), the Institute of Cost and Management Accountants of Pakistan (ICMAP) and the Management Association of Pakistan (MAP) can also play a supportive role in identifying professionals with suitable credentials to serve as independent non-executive directors. There must be some criteria established regarding the Certification of the Directors by some of the authorized institutes.
- Proper training should be incorporated for INED's when they join the company/organization. Induction should comprise of good orientation programs for directors especially INEDs which will enable them to get better acquainted with the organization, including changes in the legal, corporate, financial and other business areas, so they are able to serve their boards effectively and meaningfully. It has also been proposed by some professionals, that a minimum number of hours of training (say 10 hours) on

issues such as governance, internal control, risk management, specific courses on the business of the company etc. should be made compulsory.

- Holding of Annual General Meetings (AGMs) is one of the important forums for shareholders to assert their rights. According to SECP Annual Reports, the level of compliance with the timeliness in holding AGMs is quite high. This trend is increasing steadily, which is a positive sign. Thus holding of AGMs should be made compulsory all over the corporate sector. The strict rule should be included in the code regarding record keeping of the director's duties, board meetings, committees and their members, attendance in meetings etc. The code should make mention of the need for the board to set up different committees where required, such as the remuneration committee, audit committee and so on. Newly appointed directors and employees should be made to sign the Company's Business Ethics statement at the time of their appointment.
- The fees charged by the external auditors is also another big issue. The fee must be capped at some reasonable amount. The audit committee should comprise of the Finance literate members. Family members of the CFO, or CEO should not be the part of the Audit Committee. Once the number of family members on the Board is restricted, only then would discussions at board meetings become meaningful and transparent and the Audit Committees would be able to play a significant role.
- There must be some strict evaluation criteria for the board's performance and this should be mentioned in the code, including the person authorized to do the evaluation. Eligibility criteria for the directors should be better clarified. There is a rule regarding the vacancy filled by the board but there should be some regarding the profiles of authorized

directors so that vacancy filling process will not take much time. Number of board meetings should be relevant to the size of business.

10.4 Steps for Improving the Standards of Corporate Governance in Pakistan

The irregularities in governance could be measured through compliance with the code of corporate governance and its disclosure, which is essential for the companies operating within the country. Bauwhede (2009) presents evidence of improved performance of companies as a result of good governance. Non-compliance with the code presents the opportunity to investigate any problems and provides the opportunity for improvement. With increasing non compliance, the government of Pakistan has taken additional steps to improve corporate governance. The most important initiative taken at the time was the establishment of the Securities and Exchange Commission of Pakistan (SECP) for the purpose of developing modern and efficient laws, rules and regulations in order to protect the investors as well as shareholders from any illegal acts. The SECP started its operations on January 1, 1999 and introduced the Code of Corporate Governance (CCG) on March 28, 2002. The Code of Corporate Governance is based on internationally recognized principles and lays emphasis on openness, transparency, and accountability in the affairs of listed companies. This code of corporate governance was modified by SECP on 3rd of May 2013, and presented to comply with the Public-Sector Companies (Corporate Governance) Rules, 2013 issued for establishing a framework of good governance, whereby a public-sector company is managed in compliance with the best practices of public sector governance. In order to establish a system whereby a company is directed and controlled by its directors in compliance with the best practices so as to safeguard the interests of diversified stakeholders, it requires directors to discharge their fiduciary responsibilities diligently in the larger interests of all stakeholders. Thus it is the ultimate responsibility of the SECP to take these steps for better governance in the country:

- The responsibility of the chairman is to manage the board of directors. The chairman is the powerful person in the company. Good corporate governance revolves around his shoulders. His strong decision making can play a vital role to running the company effectively.
- There should be equal consideration of all the stakeholders whether they are minority shareholders or majority shareholders. Thus it is clear that all the shareholders are of equal importance to the company and we should consider them all in the strategic decision making for the company.
- Skills, experience and courage play vital roles if anything goes wrong. Therefore they have to make their directors aware of the current business environment.
- As the INEDs are members of the board of directors without executive responsibilities in the companies, they do judgments on the basis of their experience. They bring judgment and experience to the deliberation of the board and are supposed to identify with the interest of shareholders as well as operate in the best interest of the company in an unbiased way.
- Directors' remuneration is set by the Remuneration committee. The committee decides the remuneration of the executive and non-executive directors.
- The relationship between the auditors of the company and the client should not be cosy, but professional. External auditors should have authority and free access to information.
- The code of ethics is the set of rules and guidelines for moral behaviour and explains the guidelines for employees at work. The code also strengthens the role of directors and plays a vital role in setting the duties of the board. For this

the Pakistan Institute of Corporate Governance (PICG) should work to develop a set of detailed guidelines for board members to clearly define their duties, and boards should comply with the code.

- Corporate governance index is a tool to enable stakeholders gauge the state of corporate governance in individual companies and, overall, in a country. It indicates assessment of corporate governance practices and policies and reflects the relative level to which a company follows the generally accepted codes of corporate governance. Governance practices are measured through a survey of each company. Based on the survey findings an index is formulated. The purpose is to benchmark a company's investor right protection, disclosure quality, board effectiveness and other key features against the principles of transparency, accountability and responsibility. Different aspects of corporate governance are included in the index, such as board of directors, non-executive directors, board committees, internal audit, financial transparency and information disclosure, and stakeholders' rights.

10.5 Solutions adopted by the developed countries are not likely to be effective in the Pakistani milieu

Corporate Governance is not as effective in Pakistan as in the developed countries. The main reason for this ineffectiveness is the nature of the ownership of the companies. In the developed countries the owners of the companies are dispersed and have very minimum fraction of shares. Following the realization of the importance of corporate governance companies adopted the culture of corporate governance and began to position themselves nationally and globally. Although this was initially challenging for Pakistani companies, nevertheless, many companies strived to attain high standards of corporate governance. The major causes why many companies in Pakistan could not attain such high standards of governance are highlighted below:

- Inequitable treatment of Shareholders: In Pakistan businesses are family oriented and the whole power lies within the governing family. The minority shareholders do not have any right to participate in the decisions and they also do not unite to elect any nominee on the board for the protection of their rights and interests. The owners or the majority shareholders always go for those decisions which are directly in their favour. Inactivity or the lack of unity amongst the minority shareholders and the negligible interest towards the matters of the company is the root cause of this discrepancy.
- Insecurity of Interests of other Stakeholders : Stakeholders other than the shareholders are not as powerful to protect their interests in a developing country like Pakistan. Employees work for long hours on the minimum wage, suppliers and the customers are not satisfied with the behaviours of the companies especially as it relates to the pollution of the environment. The government is not seen to be playing an active role in addressing pollution in the environment, and this is a source of frustration for many stakeholders. These are some of the main reasons for the failure of Corporate Governance in Pakistan.
- Irresponsible Boards : Shareholders having the maximum percentage of shares are assumed to have voting rights and the selection of members of the Board of Directors. These shareholders are in the closed group and always elect people on the board who will work in their best interest. So while they appear to be fulfilling a requirement of the law, nonetheless, the benefits of having independent members on the board is not realised. For example, the requirement to

have at least one third of the board of directors to be independent is not necessarily the case.

- Unethical and Irresponsible Behaviours: Some companies carry out unethical business practices, and behave in a manner that can be regarded as not being socially responsible. This contrasts with the developed world where there is a lot of emphasis placed on doing business ethically and responsibly. Companies in Pakistan are more concerned with minimising their costs and increasing their profits, irrespective of the consequences of their actions on the environment and the local community.
- Lack of transparency in the Financial Disclosures: Companies do not always present their true financial position, which happens when internal and external auditors fail to perform their duties professionally.

When companies resolve to act ethically and do business in a responsible manner, this will be the first step toward the culture of corporate governance.

REFERENCES

Ameer, B. (2013). Corporate Governance issues and challenges in Pakistan. International Journal of Academic Research in Business and Social Sciences 3(4) ISSN: 2222-6990.

Butt, S. A. (2008), Causes of poor situation in Pakistani Listed Companies, Pakistan and

Gulf Economist.

Wenger, E. & Kaserer, C. (1998), German banks and corporate governance: a critical view, in K J Hopt, H Kanda, M J Roe, E Wymeersch, S Prigge (eds.) Comparative Corporate

Governance: the State of the Art and Emerging Research. Oxford, pp 499 – 536.

Becht, Marco, Patrick Bolton, and Ailsa Roell, 2003, Corporate Governance and Control, in Handbook of the Economics of Finance, ed. by G.M. Constantinides, M. Harris, and R. M. Stulz, Vol. 1, pp. 1–109 (the Hague: Elsevier, North-Holland).

Berglof, Erik, and Stijn Claessens, 2006, Enforcement and Good Corporate Governance in Developing Countries and Transition Economies, The World Bank Research Observer, Vol.21, No.1, pp.121–50.

Claessens, Stijn, and Joseph Fan, 2003, Corporate Governance in Asia: A Survey, (unpublished; University of Amsterdam)

"Desirable Corporate Governance: A Code", established in April 1998.15. www.k4d.org/Health/sustainable-development-challenges-and-csr-activities-in-india. [accessed 22/09/17]

Kotler, D. A. & Swedloff, R. (2003), Sarbanes-Oxley's impact on state corporate governance, Legal Opinion Letter. 13(16).

OECD, 2003, White Paper on Corporate Governance in Asia. (Paris: Organization for Economic Co-operation and Development).

Sinclair S., Campbell A., (2009), Using the crises to create better boards, http://www.mckinseyquarterly.com/Governance/Boards/Using_the_crisis_to_create_better_boards_2423# [accessed 07/09/2017]

Survey of Corporate Governance Practices in Pakistan, (2007), http://www.picg.org.pk/admin/uploadreport/Survey.pdf [accessed 10/09/2017]

Sites Used:

http://www.nation.com.pk/pakistan-news-newspaper-daily-english-online-businesscorporate-governance-ensures management accountability. [accessed 15/09/2017]

www.wikipedia.com [accessed 15/09/2017]

www.picg.org.pk [accessed 15/09/2017]

www.gcgf.org/ [accessed 15/09/2017]

issues.tigweb.org/ [accessed 15/09/2017]

www.ihrb.org/news/2011/Governance [accessed 15/09/2017]

http://www.chillibreeze.com/articles_various/CSR-in-pakistan.asp. [accessed 15/09/2017]

http://www.iisd.org/business/issues. [accessed 15/09/2017]

Chapter 11

Corporate Governance in Rwanda

Jimmy Walabyeki

11.0 Introduction

Rwanda, also referred to as "the land of a thousand hills" is a small country of about 26,000 square kilometres situated in East Africa with a population of 11.3 million people[99]. It is a landlocked country with Uganda to the north, Congo to the west, Burundi to the south and the republic of Tanzania to the East. It gained independence from Belgium on 1st July 1962 and it applied to join the Commonwealth in 2008 and was admitted to the body in November 2009, becoming the second country that was not a former colony of Britain [100] after Mozambique, to join the group of 54 countries.

The country has suffered political upheavals in the past, including the tumultuous war that ravaged the country from 1990-1994 culminating in the genocide that saw the slaughter of over 800,000 Tutsis and moderate Hutus. Since then, under the current government the country has experienced remarkable economic growth. For example, in 2016, Standard & Poor's (S&P) published an updated rating of Rwanda, confirming Rwanda's rating at "B+/B" on the lon[i]g- and short-term foreign and local currency sovereign credit[ii], which the credit rating agency attributed to "modest general government debt levels, government's commitment to fundamental reforms; including in terms of ease of doing business, strong economic growth, all supported by strong investment performance and strengthening exports and flexible exchange-rate regime"[101] with a monetary policy that controls inflation.

Indications are that these economic developments have had a trickle-down effect on Rwandans, resulting in social development. This is shown by the Human Development Index which highlights the fact that Rwanda has been steadily rising from the Low Human Development level of 0.244 in 1990, rising to 0.498 in 2016, with current trends showing that Rwanda shall soon reach Medium Human development category.[102]Also known as the Singapore of Africa, for its excellent economic and social reforms, it has become a beacon of governance on the continent especially with respect to good institutions, ethics and a good business environment. These achievements have been characterized by significant improvements in corporate governance. In the past good corporate governance was not a major priority in Rwanda, as the economy was characterized by a small private sector, dominance of the state in business, small family owned businesses, and no capital markets.[103] However, several factors, both direct and indirect, have contributed to improved corporate governance in Rwanda. These factors will be the subject of this article commencing with the direct factors.

11.1 Direct Factors

Direct factors include deliberate efforts by the government or other entities to improve corporate governance. It is worth noting that since joining the Commonwealth of Nations in 2008, there has been some foreign direct investment into Rwanda's economy. However, there has been no direct significant effect on corporate governance by Rwanda's new involvement with the organization. However, there has been direct external influence from other international organizations including the Bretton woods institutions namely the World Bank and the International Monetary Fund. Other institutions involved include the East African community, the African Union, the United Nations, the Organization for Economic Co-operation and Development *(OECD)* and the International Organization of Securities Commissions (IOSCO) among others. The discussion will commence with governance reviews.

11.2 Governance Reviews

Internal and external reviews have made a significant contribution to the development of good corporate governance in Rwanda. First, they have been instrumental in identifying corporate governance weaknesses, thereby providing a basis for recommended solutions. Second, reviews have enabled policy makers to monitor the effectiveness of the reforms or measures taken to improve corporate governance. External reviews include the Africa Peer Review Mechanism in 2005[104] under the umbrella of the African Union, and the Report on the Observance of Standards and Codes (ROSC) report in 2008[105] carried out under the auspices of the International Monetary fund and World Bank.

Internally, unique to Rwanda, there is an annual review or audit of governance, including corporate governance, by the Rwanda Governance Board, where the state of governance, including corporate governance, is reviewed every year.[106] In addition, there is the National Leadership Retreat (*Umwiherero*) which is an annual leadership retreat bringing together senior government officials of the Government of Rwanda [107], chaired by the president, to address the challenges facing the country, and is organized by the Office of the President, and the Office of the Prime Minister. At the retreat, a broad range of issues, including economics, politics, justice, infrastructure, health, education and governance, including corporate governance, are discussed. Since its inception, organisers of Umwiherero have adopted numerous initiatives to improve the implementation of resolutions agreed upon at each retreat. By 2011, these efforts resulted in noticeable improvements in planning, coordination, and accountability leading to clearer and more concise priorities. In 2011 six priorities were identified down from 174 in 2009, allowing for more effective delivery and implementation of Umwiherero resolutions. Reviews are very effective in Rwanda because there is a strong political will to implement the recommendations arising from the reviews. For example, it has been noted that Rwanda has from World Bank's *Doing Business 2005* to *Doing Business 2017* implemented a total of 47 reforms across all indicators. It has also been identified as one of only 10 economies that have implemented reforms in all of the *Doing Business* indicators and every year since *Doing Business 2006*.[108] In addition, in the latest ROSC Report (for Rwanda) of 2015 it was noted commendably that of the 21 recommendations made in the ROSC report of 2008, seven had been fully implemented, with 11 partially implemented, and 3 had not yet been implemented. In keeping with this trend, there have been significant regulatory reforms which have taken place following governance reviews including enactment of legislation and codes, and establishment of institutions, both of which shall be considered as follows.

11.3 Legislation and Codes

These have provided the right legal frame work for good corporate governance. In this regard, Rwanda has ratified and adopted a significant number of key international standards and codes in corporate governance[109] including the New Partnership for Africa's Development (NEPAD) Framework Document (2001), Principles of Corporate Governance (OECD and Commonwealth), International Accounting and Auditing Standards, Core Principles on Effective Bank Supervision, African Charter on Human and People's Rights, Labour Codes of the International Labour Organisation, and Codes on Industrial and Environmental Safety and Hygiene of the World Health Organisation. [110]For example, concerning Health and Safety, Rwanda has an Occupational Health policy pursuant to these codes and conventions as the policy states that "Rwanda's Occupational Safety and Health (OSH) concerns mirror International trends that seek to improve Safety and Health in the work environment as defined by the International Labour Organization. Rwanda has therefore aligned its OSH response to ILO Convention 187 (2006) on the Promotional Framework for Occupational Safety and Health and Convention 155 on Occupational Safety and Health and the working environment (1981)"[111]. Further, pursuant to these international codes and conventions over the last 10 years the government of Rwanda has passed several laws, and some voluntary codes which reflect international principles of corporate governance.[112] What follows is a discussion of the key local laws and codes relevant to corporate governance.

11.3.1 The Companies Act, 2017

The Companies Act replaced the Companies Act, 2009 and gives the foundational corporate governance framework providing for among others, incorporation of companies, types of companies, duties of directors, rights of shareholders, protection of minority shareholders, maintenance of capital and reporting requirements such as audits and filing annual returns.

11.3.2 Financial Sector Laws

The National Bank of Rwanda is governed by *Law No. 55 of 2007 governing the Central Bank. Law No. 007 of 2008 concerning the organization of banking* deals specifically with other Banks including commercial Banks. There are other laws such as *Law No. 47 of 2008 on Money Laundering and Financing Terrorism* which cover money laundering and terrorism financing and related criminal activity. There are also regulations concerning insider lending, credit classification, and corporate governance. *Corporate governance regulation (No. 06/2008) for financial institutions* provides for duties of shareholders, requirements for approval of board members, internal audit requirements, conduct of board meetings, board committees, responsibilities of management and a code of conduct. These laws were enacted not only pursuant to global trends, but because the Rwandan economy was plagued by a spate of corporate collapses in the Banking sector.[113] Of note was the collapse of *The Bank of Commerce, and Development and Industry (BCDI)* and the *Banque Continentale Africaine* (BACAR).[114] Poor corporate governance practices were a major cause of these corporate failures. For insurance companies, *Law N°07/2009 of 29/07/2009* establishes duties, responsibilities and a code of conduct for shareholders, directors, and management of licensed insurers or brokers, and there is also *Corporate governance regulation (No.07/2009)* covering companies in the insurance sector. Other financial institutions are covered by the *Pension Law (No.05/2016)*, Micro Finance Institutions are governed by *Law (No.40/2008) and Regulations on Micro Finance Institutions*. All of the above laws governing financial institutions are based on international best practices like Basel I & II, OECD Principles, and King II. [115]

11.3.3 Securities and Capital Markets Laws

The *Law N°11/2011 of 18/05/2011 Establishing the Capital Market Authority (CMA) and Determining its Mission, Powers, Organisation and Functioning* establishes the Capital Market Authority. There are other regulations covering Capital markets including *The Capital Market Corporate Governance Code No. 9, 2012* which provides important provisions for corporate governance, also requiring education of shareholders on their rights, and the need to educate the media on the principles of corporate governance.The code applies to all listed companies, and public companies are strongly encouraged to comply with the provisions of the code, and private companies are also encouraged to comply.[116]With respect to the enforcement of the laws, the Capital Markets Authority adopts a comply or else approach, but also expressly provides that "[w]here a firm or other person has failed to comply with the requirements of the Act, the regulations, or other relevant legislation, it may be appropriate to deal with this without the need for formal disciplinary or other enforcement action"[117].

11.3.4 The Private Sector Federation Corporate Governance Code

There is no all embracing code for corporate governance in Rwanda, but the Rwanda Private Sector Federation (PSF) produced arguably the most far reaching code of corporate governance in Rwanda. In 2009 it launched a code of business ethics, and a code of corporate governance in 2009, to promote integrity, transparency and professionalism in the business community in Rwanda. The code applies to the directors and managers of member companies, non-executive directors, shareholders, and all employees of the company. Provisions include reporting requirements, board committees, and duties of directors among others.

11.3.5 Laws protecting stakeholders

Concerning stakeholders, great strides have been made to protect their interests within the corporate governance matrix. For example labour laws have been enacted to protect employees[118], and there must be compliance with environment laws[119] by both local and foreign investors. A major highlight in the protection of stakeholders is the constitutional provision requiring women to hold at least 30 percent of the positions in decision making organs[120], which may have an impact on boards, ensuring gender sensitivity on board appointments. To protect consumers, Rwanda has a Competition and Consumer Protection policy whose purpose is "to promote fair competition, providing consumers access to products and services at competitive prices and better quality and to create an environment, which is conducive to investment"[121]. The relevant law in this respect is the *Law N° 36/2012 OF 21/09/2012 Relating to Competition and Consumer Protection.* In all, Rwanda has adopted a hybrid system of regulation of corporate governance with compulsory legal provisions, and voluntary codes. To ensure compliance with these regulations, Rwanda has established institutions, and a commercial justice system which shall be discussed hereafter.

11.4 Building Institutions

In the arena of institution building, Rwanda has made remarkable progress, as shown by the country's ranking 1st in institution establishment in sub – Saharan Africa in the 2016-2017 Global Competitive Index (GCI) Report of the World Economic Forum.[122]The key institutions relevant to corporate governance shall be considered in turn.

11.4.1 The National Bank of Rwanda (BNR)

The National Bank of Rwanda is established by law No 55 of 2007 and its functions include maintaining price stability, maintaining a stable and competitive financial system, and supporting general economic policies of the government. The BNR provides supervisory oversight over commercial banks, microfinance institutions, forex bureaus and non-bank financial institutions including pensions and insurance.

11.4.2 The Capital Markets Authority

Capital Markets in Rwanda are regulated by the Capital Market Authority (CMA)[123], and its duty among others, is to implement the Government policy on Capital Markets, prepare draft policies on capital markets, and promote fairness, efficiency and transparency in the functioning of the capital market. It also protects Rwandan citizens and investors in capital markets from unfair and unsound practices or practices involving fraud, deceit, cheating or manipulation. The Capital Market Authority is an associate member of the International Organization of Securities Commissions (IOSCO), and so the Authority benefits from the knowledge which IOSCO shares with its members.

11.4.3 Rwanda Development Board

The Rwanda Development Board was established by merging all the government agencies responsible for facilitating both local and foreign investment in one place, thereby establishing a one stop shop for investment. [124]The Board reports directly to the President, and at the top of its governance structure is a Board of Directors which includes all the ministries relevant for investment like finance, commerce, infrastructure, and agriculture. The Development Board draws technical assistance and advice directly from countries with greater experience in investment and corporate governance, like Singapore and Costa Rica. It also has direct support from global entrepreneurs and experts from Singapore Development Board, World Bank, IFC and the Office of Tony Blair [125]. Not surprisingly, Rwanda ranked 56th in the latest Doing Business Report 2017, the second in Africa as regards the ease of doing business, as at June 2016.[126]

11.4.4 Rwanda Governance Board

The Rwanda Governance Board (RGB)[127] regulates Faith Based Organizations (FBO), Non-Governmental Organizations (NGO's), and Civil Society Organizations (CSO), ensuring that they conform with their respective regulatory requirements.

11.4.5 Government Portfolio Management Unit (GPMU)

Relevant to corporate governance of state owned enterprises, the Government Portfolio Management Unit [128] is a department that operates under the auspices of the Ministry of Finance and

Economic Planning, whose stated purpose is to ensure that Government Investments are managed to achieve the Government's strategic objectives.[129] It ensures that they comply with legislation and operate profitably. The unit promotes corporate governance and risk management in Government Business Enterprises (GBEs) by "strengthening financial and accountability discipline within public enterprises, and ensuring optimum government investment portfolio to enhance sustainable economic development"[130]. It is commendable that there is a special unit dedicated to ensuring good corporate governance in government institutions, however, it is important to note that Rwanda does not ascribe to the OECD code of corporate governance for state owned enterprises. [131]

11.4.6 Consumer Protection Institutions

A number of institutions exist specifically to ensure protection of consumers. These include Rwanda Utilities Regulatory Authority (RURA)[132] which regulates public utilities, like telecommunications services, electricity, water, removal of waste products from residential or business premises, extraction and distribution of gas and transport of goods and persons. It links policymakers, licensed service providers and consumers in fostering fair competition, and protecting consumers' interests. Other institutions include the Competition and Consumer protection Unit under the Ministry of Trade and Industry (MINICOM) which "provides consumers with competitive prices and product choices at the best possible quality"[133] , and enhances "the competitiveness of Rwandan enterprises in world markets by exposing them to competition within the country"[134]. Also, the Association for the Defense of Consumer Rights (ADECOR) which is a civil society organisation campaigns and lobbies for consumer rights, gives consumer advice, and provides dispute resolution between consumers and providers of goods and services[135]. Finally, the Rwanda Bureau of Standards (RBS) protects consumers and promotes trade within and outside the country, and gives industrial certification to confirm that products meet required standards[136].

These institutions highlight the importance of the protection of stakeholders like consumers, in a good corporate governance framework. The investment code act regulates local investment both by nationals and foreigners ensuring an effective regime not only for protection of foreign direct investment interests[137] but also regulation of local investors.

11.4.7 Professional Bodies

These organizations regulate and set standards of conduct for their professions and in some cases enforce ethical standards. For accountants and auditors, the Institute of Certified Public Accountants of Rwanda (ICPAR) was created, and empowered by *Law number 11/2008 to regulate the Accounting profession in the republic of Rwanda*[138] , and set minimum qualification requirements for accountants in Rwanda. Lawyers are regulated by the Rwanda Bar Association, and it "provides legal information, advice, orientation, mediation, and legal assistance & representation to indigent and vulnerable people free of charge". It also provides training programmes for Advocates, including training in corporate governance. [139] For bankers, the Rwanda Bankers Association was established to promote the standards of professional competence and practice among its members. [140] For the general business community in Rwanda, the Private Sector Federation - Rwanda (PSF) is a professional organization whose goal is to represent the members' goals and aspirations. The PSF is an umbrella organization that groups 10 professional chambers, including the Chambers of Agriculture, Commerce, Art & Crafts, Finance, Information and Communications Technology (ICT) Industry, Liberal Professions, Tourism, Women Entrepreneurs, and Young Entrepreneurs. The PSF spearheaded the Business Development Services (BDS) which contributes to sensitization of corporate governance practices and promoting corporate performance.[141]Other professional bodies include the Association of Micro- Finance Institutions in Rwanda, the Rwanda Journalists Association, and the Rwanda Miners' Association among others. It is important to note that such professional bodies typically lack the key ingredient of independence from the same professionals, whom they oversee.[142] It is not clear whether the professional bodies in Rwanda have overcome this potential handicap, but nevertheless they play an important role in contributing to the regulation of their respective members to ensure good corporate governance. Another challenge faced by these professional bodies is that there is inadequate technical and financial capacity. For example, the ICPAR, faces the problem of technical capacity because of a shortage of qualified professional accountants in Rwanda, and because it has been hard to attract and retain qualified staff. [143]

11.4.8 Rwanda Economic and Social Development Council

In order to ensure that there are concerted efforts that bring together the interests of different institutions and players reflected by the above institutions, by Presidential Order N° 64/01 of 31/12/2007 on *Establishment and Functioning of Rwanda Economic and Social Development Council*, this institution was formed. Its purpose is to place a framework for improved dialogue and enhanced consensus building of "nationally owned, integrated and evidence-based development policy making between Government, Business Community and Civil Society actors in Rwanda with the aim of improving the social and economic development processes of the country". This organization has contributed greatly to promoting the successes that Rwanda has so far experienced in achieving its Millennium Development goals.[144] To ensure that laws are enforced, another mechanism in place in Rwanda is a commercial justice system.

11.5 Dispute Settlement

To ensure that commercial disputes are handled expeditiously, there is a commercial court which has a country wide reach[145], and this has greatly contributed to the business environment in the country.[146] The strides taken in this respect have provided a favorable environment for the protection of minority shareholders, and foreign investments in Rwanda.[147]As an alternative to the adversarial court system Rwanda has an international Centre for arbitration called the Kigali International Arbitration Centre, where mediations and Arbitrations can be carried out.[148] This has the effect of enabling businesses to preserve business relationships, helping conflicting parties to resolve their disputes expeditiously.[149] It is not enough to have an effective regulatory system on corporate governance without sensitizing Rwandans about these principles. For this reason, there have been deliberate efforts to ensure that the principles of corporate governance are disseminated to relevant stakeholders.

11.6 Capacity Building Programs

Both local and international organizations or development partners have been involved in capacity building, and training in corporate governance. For example the International Finance Corporation (IFC) has been involved in promoting corporate governance amongst SMEs in Rwanda. In this respect it established an online training program and a business solutions center, that is geared towards helping Rwandan entrepreneurs to develop new skills and build stronger businesses, placing emphasis on good corporate governance practices. The online training program offers guidance on different topics including marketing productivity and accounting.[150] As regards listed companies, the IFC has facilitated training in relevant areas of corporate governance tailor made for companies whose shares are traded on the Rwanda Stock Exchange.[151] There have also been collaborative efforts between the government and its development partners in order to promote knowledge of good corporate governance principles. For example, by 2005 the Government and the Private Sector Federation held sensitisation workshops with the assistance of the World Bank-funded Competitiveness and Enterprise Development Project, in an effort to raise awareness of corporate governance in both the private and public sectors.[152] Local Organizations providing training in corporate governance include Lion Corporate Solutions, Institute of Certified Public Accountants of Rwanda, University of Kigali Centre for Economic Governance and Leadership, Private Sector Federation of Rwanda, and the soon to be formed Rwanda Institute Of Directors (RIOD). A lot more needs to be done in respect of capacity building, including the formation of RIOD, whose operationalization is greatly anticipated.

11.7 Indirect Factors

Indirect factors are efforts including government policies, which are put in place, not directly to improve corporate governance, but when they are implemented, have a positive effect on corporate governance.

11.7.1 Economic Reforms

Following global economic trends, the government of Rwanda departing from previous policies, has first, adopted a more market based economy, the effect of which has been to place greater emphasis on companies, and on how they are run efficiently thereby highlighting the need for corporate governance.[153] This has increasingly made good corporate governance to take centre stage. [154] Second, the establishment of the Kigali Securities Exchange in March 2007 has enhanced the standards of corporate governance in Rwanda, because listed companies are required to conform with the basic corporate governance principles.[155]So far, Rwanda has one stock exchange whose stated purpose is "to provide a platform for companies to raise capital for their businesses, to provide an avenue for Rwandans to mobilize savings for investment, to promote good corporate governance for companies, enable the government to raise capital for investment projects, and to be a barometer for the economy"[156]. Third, there was a deliberate policy of privatization of government parastatals. Privatization 'helps to promote corporate governance because it highlights corporate governance concerns in sectors that were previously owned and controlled by the state, but where now private owners go to the market to seek financing' [157]. In keeping with global trends, the goal of privatization was "to relieve the financial and administrative burden of government, improve efficiency and productivity of these companies, and to reduce the size and presence of the public sector in the economy"[158]. In Rwanda, the privatization process started in 1996 with the main aim of creating a dynamic, export-oriented economy driven by free enterprise, and since then, at least 56 state owned enterprises have been privatized. Finally, the drive to attract foreign direct investment has had the effect of enhancing corporate governance in Rwanda. [159] The impact of foreign direct investment has resulted in a direct implementation of good standards of corporate governance in the country imported from the foreign companies' jurisdictions. For example the United States State department has observed that the awareness of corporate social responsibility in Rwanda is growing, but it is mainly foreign owned companies which have implemented sustainable programs of the same.[160]This means that some good practices relating to good corporate citizenship have been introduced or enhanced within Rwanda's jurisdiction, which contributes to a culture of corporate social responsibility amongst companies in Rwanda. Another important factor is Rwanda's fight against corruption.

11.7.2 Anti-Corruption Policy

A strong stand against corruption in Rwanda has created a good environment for high standards of corporate ethics. Rwanda is a signatory to the United Nations Convention Against corruption (UNCAC), and in 2012 she launched her anti-corruption policy, whose goal is to promote and inculcate awareness of a society founded on discipline, integrity and ethics, and to suppress corruption through effective means.[161] Institutions that are involved in the fight against corruption include the National Public Prosecution Authority, the Rwanda National Police, the Office of the Auditor General of State Finances, Rwanda Public Procurement Authority, Parliamentary Public Accounts Committee and the Courts.[162]Also, to create a framework of exchange of information on corruption and consolidate the institutions' efforts to combat corruption, the National Anti-Corruption Advisory Council was set up.[163] All in all, Rwanda has made great strides in the fight against corruption, as shown by the fact that on Transparency International's Perceptions of corruptions index 2016 Rwanda ranked 50th second to Botswana in Africa. However, challenges still face the fight against corruption including the citizens' limited involvement in reporting corruption, and resource challenges facing institutions in charge of fighting corruption.[164] For example, Rwanda needs a forensic laboratory at the Office of the Ombudsman. There is also a greater need to engage civil society groups and the media in the fight against corruption.[165] Finally, by 2012 it was noted that some provisions like international cooperation and mutual legal assistance are still missing in the Rwandan legislation[166], but to this author's knowledge, these *lacunae* have not yet been filled.

11.8 Conclusion

The future of Rwanda is very bright, and because of good corporate governance among other reasons the nation is set to attract more local and foreign investment. Arthur Levitt said "If a country does not have a reputation for strong corporate governance practices, capital will flow elsewhere"[167]. However, from the above it is manifest that Rwanda has a good system of corporate governance, which following recent performance is set to get better. It is therefore projected that it shall increasingly be an attractive destination for investors. The development of corporate governance in Rwanda has been greatly influenced by internal players in Rwanda, both public and private, as well as her development partners including other countries and international bodies. Indeed, for the trends to continue, the same concerted effort will have to continue to take the corporate governance standards to the next level.

REFERENCES

African Peer Review Mechanism, (2005), African Union Al Zuhair, M., (2003), "Privatization Programs, Ownership Structures, and Market Development: The Role of Country Characteristics in Defining Corporate Governance Standards", available at

http://pqdtopen.proquest.com/doc/193988520.html?FMT=AI [accessed 20/03/2017].

Barenzi, R., (2010), "Understanding Corporate Social Responsibility" *New Times* Newspaper, available at http://www.newtimes.co.rw/section/article/2010-09-28/24408/

Hertveldt, S., (2008), "Rwanda: Pragmatism leads the way in setting up specialized commercial courts", available at http://www.doingbusiness.org/reports/case-studies/2008/building-specialized-commercial-courts-in-rwanda [accessed 22/02/2017].

Johnson, S. and Shleifer, A., (2004) "Privatization and Corporate Governance", p.13-14, available at

http://www.nber.org/chapters/c10182.pdf accessed on 20th March 2017 . where that author notes that where there is poor corporate governance of privatized institutions, they under perform.

Kamasa, K., (2017), "National Leadership Retreat: unique way of holding leaders accountable" *New Times* Newspaper, available at http://www.newtimes.co.rw/section/article/2017-02-26/208377/ [accessed 7/03/2017].

Kayihura M. D., (2013), "The Corporate Governance Approach in Light of classical Approaches: The Shareholder v The Stakeholder. The Case of Rwanda" *Rwanda Journal, Series H: Economics and Management Vol. 1 (1)*

Kwibuka, E., (2016), "Ombudsman outlines plans to intensify corruption fight" *The New Times* (Rwanda) available at http://www.newtimes.co.rw/section/article/2016-02-22/197313/ [accessed 27/03/2017].

Malunda, D., (2015), "The State of Play of Competition Policy and Law Reforms: The Case of Rwanda", available at http://www.cuts-international.org/ARC/Nairobi/EACCOMP/pdf/The_State_of_Play_of_Competition_Policy_and_Law_Reforms-The_Case_of_Rwanda.pdf [accessed 25/03/2017].

Ministry of Finance and Economic Planning, "Standard & Poor's confirms Rwanda's rating at B+ Outlook Negative" http://minecofin.gov.rw/index.php?id=12&no_cache=1&tx_ttnews%5Btt_news%5D=490 accessed on the 7th March 2017.

Ministry of Public Service and Labour, Occupational Safety and Health Policy

http://www.mifotra.gov.rw/fileadmin/user_upload/labour_pictures/OSH_POLICY.pdf accessed on the 14th March 2017 .

Morara, B., (June 2010), "Enhancement of Corporate Governance Through Privatization", Privatization Commission, Kenya.

Mwika, J., (2012), "Is it vital to have Corporate Governance Codes for Institutional Investors & Capital Markets? A Case Study of Rwanda", Karlstad Business School, Sweden.

OECD, "Alternative Dispute Resolution and Corporate Governance" available at

http://www.oecd.org/daf/ca/corporategovernanceprinciples/alternativedisputeresolutionandcorporategovernance.htm

Office of the Ombudsman, (2012), "Rwanda Anti- Corruption Policy": p.6

http://www.minijust.gov.rw/fileadmin/Documents/Policies/Rwanda_Anti_Corruption_Policy.pdf [accessed 27/03/2017].

Pflanz, M., (2009), "Rwanda joins the Commonwealth" http://www.telegraph.co.uk/news/worldnews/africaandindianocean/rwanda/6685316/Rwanda-joins-the-Commonwealth.html Accessed on the 2nd February 2017.

Rwanda Joint Governance Assessment Report (2008)

Rwangombwa, J., (2016), "Improving Corporate Governance : The Rwandan Financial Sector Perspective", available at http://events.afreximbank.com/wp-content/uploads/2016/08/Session-5-Improving-Corporate-Governane-Practices-The-Rwanda-FI-Perspective.pdf [accessed 27/03/2017].

Walabyeki, J., (2014), "The Role of Professionals in Corporate Governance" in *Corporate Governance in Uganda*

SMEToolkit, "Rwanda", at

http://rwanda.smetoolkit.org/rwanda/en/category/4821/Public-Relations accessed on the 7th March 2017.

United States, (2015), "State Department's Office of Investment Affairs' 2015 Investment Climate Statement." Available at https://www.export.gov/article?id=Rwanda-Corporate-Social-Responsibility accessed on the 3rd March 2017.

Walabyeki, J., (2016), "Corporate Governance in Rwanda: The Role of the External Auditor", Available at SSRN: https://ssrn.com/abstract=2760806 or

http://dx.doi.org/10.2139/ssrn.2760806

World Bank Group, (2017), Doing Business.

LEGISLATION

Capital Market Corporate Governance Code No. 9, 2012.

Constitution of Rwanda, 2015 as amended.

Law n° 39/2001 of 13 September 2001

Law N° 04/2005 of 08/04/2005 Organic Law determining the modalities of protection, conservation and promotion of Environment in Rwanda.

Law N° 26/2005 Of 17/12/2005 Relating to Investment and Export Promotion and Facilitation.

Law No 06/2009 of 21st December 2009

Law N° 13/2009 of 27/05/2009 Law regulating labour in Rwanda.

Law No 41/2011 of 30/09/2011.

Regulation 06 on Capital Market Enforcement Guidance, 2012.

WEBSITES

Capital Market Authority Rwanda website at

http://www.cma.rw/cma-background on 17th November 2016.

ICPAR website at http://www.icparwanda.com

IFC website at

http://www.ifc.org/wps/wcm/connect/news_ext_content/ifc_external_corporate_site/news+and+events/rwanda_sme_toolkit accessed on the 7th March 2017.

Ministry of Finance and Economic Planning website at http://www.minecofin.gov.rw/index.php?id=153 accessed on the 6th March 2017.

Private Sector Federation website at

http://www.psf.org.rw/spip.php?rubrique1 accessed on 25th February 2017.

Rwanda Bankers Association Website at

http://www.rba.rw/about.php accessed on the 22nd March 2017.

Rwanda Bar Association Website at

http://www.rwandabar.org.rw/service-to-advocates/ accessed on the 22nd March 2017.

Rwanda Development Board website at http://www.rdb.rw/about-rdb/vision-and-mission.html accessed on the 25th February 2017.

Rwanda Stock Exchange website at

http://www.rse.rw/site/pages/role+of+rse/5 accessed on the 17th November 2016.

Supreme Court website at

http://www.judiciary.gov.rw/the_courts/specialised_courts/commercial_high_court/ accessed on the 22nd February 2017.

Endnotes

---

99 Klaus Schwab, World Economic Forum Report (2016-2017) p. 308.

100 Mike Pflanz, 'Rwanda joins the Commonwealth' (29th November 2009)

http://www.telegraph.co.uk/news/worldnews/africaandindianocean/rwanda/6685316/Rwanda-joins-the-Commonwealth.html [accessed 2/02/2017].

[101] See Ministry of Finance and Economic Planning, 'Standard & Poor's confirms Rwanda's rating at B+ Outlook Negative' (12th March 2016) http://minecofin.gov.rw/index.php?id=12&no_cache=1&tx_ttnews%5Btt_news%5D=490 [accessed 7/03/2017].

[102] See Rosa Malango, 'UNDP's 2016 human development report enables Uganda to reflect on the social sector' New Vision Newspaper (20th March 2017) http://www.newvision.co.ug/new_vision/news/1449652/undp-2016-human-development-report-enables-uganda-reflect-social-sector [accessed 30/03/2017]. See also United Nations Development Program Human Development Report 2016: Human Development for Everyone available at http://hdr.undp.org/sites/default/files/2016_human_development_report.pdf [accessed 30/03/2017].

[103] Rwanda Joint Governance Assessment Report (12th September 2008) p.73 (A Report following a joint review between the Rwanda Government and her development partners including USA, UK, Belgium, the Netherlands, Swiss, EU and Germany, including heads of development agencies namely; the World Bank, African Development Bank (ADB), United Nations Development Program (UNDP) , United States Agency for International Development, Canadian International Development Agency (CIDA), The Department for International Development (DFID), Belgium Technical Co-operation, and the German Agency for Technical Cooperation (GTZ).

[104] See African Peer Review Mechanism (APRM), 'Country Review Report of the Republic of Rwanda' (November 2005).

[105] See The Report on the Observance of Standards and Codes - Accounting and Auditing (ROSC A&A) (November 2015) which was a follow up from the previous ROSC Report of 2008.

[106] See for example the Rwanda Governance Board, 'Rwanda Governance Scorecard, 2016 : The State of Governance in Rwanda' (December 2016).

However the corporate governance indicators are limited to the number of Small and Medium Sized enterprises registered, and their mortality rate.

[107] See Kim Kamasa, 'National Leadership Retreat: unique way of holding leaders accountable' *New Times* Newspaper (February 26, 2017) available at

http://www.newtimes.co.rw/section/article/2017-02-26/208377/ [accessed 7/03/2017].

[108] World Bank, 'Doing Business: Comparing Business Regulation in 190 Countries', 2017.

[109] APRM Report *supra* note 5, p.81.

[110] APRM Report *ibid.*

[111] See Ministry of Public Service and Labour, Occupational Safety and Health Policy p.7 http://www.mifotra.gov.rw/fileadmin/user_upload/labour_pictures/OSH_POLICY.pdf accessed on the 14th March 2017 . The legal framework is outlined on p.9 of the policy.

[112] Rwanda Joint Governance Assessment Report, *supra* note 5 p. 74.

[113] Kayihura M D, 'The Corporate Governance Approach in Light of classical Approaches: The Shareholder v The Stakeholder. The Case of Rwanda' *Rwanda Journal, Series H: Economics and Management Vol. 1 No 1, 2013* p.23 cited in Walabyeki, Jimmy, Corporate Governance in Rwanda: The Role of the External Auditor (April 8, 2016). Available at SSRN: https://ssrn.com/abstract=2760806 or

http://dx.doi.org/10.2139/ssrn.2760806

[114] *Ibid.*

[115] John Rwangombwa, 'Improving Corporate Governance : The Rwandan Financial Sector Perspective (October 2016)'available at http://events.afreximbank.com/wp-content/uploads/2016/08/Session-5-Improving-Corporate-Governane-Practices-The-Rwanda-FI-Perspective.pdf [accessed 27/03/2017].

[116] Article 3 of the *The Capital Market Corporate Governance Code No. 9, 2012.*

[117]Article 2 of the *Regulation 06 on Capital Market Enforcement Guidance, 2012.*

[118] Law N° 13/2009 of 27/05/2009 Law regulating labour in Rwanda.

[119] N° 04/2005 of 08/04/2005 Organic Law determining the modalities of protection, conservation and promotion of Environment in Rwanda.

[120] Article 10 of the Constitution of Rwanda, 2015 as amended.

[121] Ministry of Trade and Industry, 'Rwanda Competition Policy' (2010) p. 6.

[122] Klaus Schwab, World Economic Forum Report *supra* note 1 p. 38. On the same index Rwanda emerged 52nd overall scoring 4.41 after Mauritius and the Republic of South Africa respectively in Africa.

[123] See Capital Market Authority Rwanda website at http://www.cma.rw/cma-background on 17th November 2016.

[124] See Rwanda Development Board website at http://www.rdb.rw/about-rdb/vision-and-mission.html [accessed 25/02/2017].

[125] *Ibid.*

[126] See World Bank, *supra* note 9, p. 7.

[127] It is a 'public agency with legal personality, administrative and financial autonomy, established by law No 41/2011 of 30/09/2011. It is managed in accordance with Organic law No 06/2009 of 21st December 2009 establishing general provisions governing public institutions functioning and administration'. See Rwanda Governance Board website at http://www.rgb.rw/index.php?id=15 accessed on the 29th March 2017.

[128] See Ministry of Finance and Economic Planning website at http://www.minecofin.gov.rw/index.php?id=153 [accessed 6/03/2017].

[129] *Ibid*

[130] *Ibid.*

[131] Export.org, 'Rwanda - Competition from State-Owned Enterprises' available at

https://www.export.gov/apex/article2?id=Rwanda-Competition-from-State-Owned-Enterprises [accessed 27/02/2017].

[132] It was established by *Law n° 39/2001* of 13 September 2001.

[133] Dickson Malunda, 'The State of Play of Competition Policy and Law Reforms: The Case of Rwanda' (2015) p. 35 available at http://www.cuts-international.org/ARC/Nairobi/EACCOMP/pdf/The_State_of_Play_of_Competition_Policy_and_Law_Reforms-The_Case_of_Rwanda.pdf [accessed 25/03/2017.

[134] *Ibid*

[135] *Ibid*

[136] *Ibid*

[137] Law N° 26/2005 Of 17/12/2005 Relating to Investment and Export Promotion and Facilitation.

[138] See Institute of Certified Public Accountants of Rwanda (ICPAR) website at http://www.icparwanda.com

[139] See the Rwanda Bar Association Website at

http://www.rwandabar.org.rw/service-to-advocates/ accessed on the 22nd March 2017.

[140] See the Rwanda Bankers Association Website at http://www.rba.rw/about.php [accessed 22/03/2017].

[141] See Private Sector Federation website at http://www.psf.org.rw/spip.php?rubrique1 accessed on 25th February 2017.

[142] Jimmy Walabyeki, 'The Role of Professionals in Corporate Governance' in *Corporate Governance in Uganda* (2014)

[143] ROSC report supra note 6 p. 7.

[144] The Hon. Minister of Finance and Economic Planning, 'Rwanda Speech at the Operational Activities for Development Segment of ECOSOC in New York '22nd – 24th February, 2016 stated thus; 'Overall Rwanda has all the Millennium Development Goals and in some cases exceeding the targets particularly those related to; gender, education, maternal and infant health. Important factors that have contributed towards this good performance include; strong national ownership and visionary leadership, institutionalization of a results-based management culture, nurturing of home-grown citizen-centred innovations, effective monitoring mechanisms, institutionalized accountability as well as strengthened institutional and individual capacities'.

[145] See Supreme Court website at

http://www.judiciary.gov.rw/the_courts/specialised_courts/commercial_high_court/ [accessed 22/02/2017].

[146] Justice Rugege cited at the Supreme Court website at available at http://www.judiciary.gov.rw/the_courts/specialised_courts/commercial_high_court/ [accessed 22/02/2017].

[147] See OECD. Alternative Dispute Resolution and Corporate Governance available at http://www.oecd.org/daf/ca/corporategovernanceprinciples/alternativedisputeresolutionandcorporategovernance.htm accessed on the 27th March 2017 where it is stated that 'The Corporate Governance Regional Roundtables have identified poor regulatory and judicial enforcement as a significant impediment to shareholder protection and foreign investment in emerging markets. Deficiencies in emerging market enforcement have helped make arbitration of company law disputes (ACLD) the preferred method of dispute resolution for private-equity investors in many key emerging markets, such as Russia and China'.

[148] See also Sabine Hertveldt, 'Rwanda: Pragmatism leads the way in setting up specialized commercial courts' 2008 available at http://www.doingbusiness.org/reports/case-studies/2008/building-specialized-commercial-courts-in-rwanda accessed on the 22nd February 2017.

[149] King III states that 'It is suggested that alternative dispute resolution will enable business to preserve business relationships, by speedily solving problems'.

[150] The online training tools and business centre were launched by IFC's Rwanda Entrepreneurship Development Program (REDP), a two-year, $2.4 million geared towards supporting the growth of smaller businesses in Rwanda with training and tools to bridge entrepreneurship development skills gaps. See IFC website at

http://www.ifc.org/wps/wcm/connect/news_ext_content/ifc_external_corporate_site/news+and+events/rwanda_sme_toolkit [accessed 7/03/2017]. See also the SMEToolkit Rwanda at

http://rwanda.smetoolkit.org/rwanda/en/category/4821/Public-Relations [accessed 7/03/2017] to see more information about this online resource.

[151] For example, in December 2016, IFC conducted a workshop for listed companies, and at the workshop, The Executive Director of Capital Market Authority (CMA), Robert Mathu stressed that 'IFC has consistently cooperated with CMA in helping companies in Rwanda to explore capital markets as a source of long-term finance'. See Capital Market Authority website at available at http://www.cma.rw/content/cma-and-ifc-improving-corporate-governance-practices-rwanda [accessed 7/03/2017].

[152] APRM p.82. Also noteworthy in this respect was the Multi-Sector Capacity-Building Programme (MSCBP), which was a long-term government strategic framework geared towards capacity building in the public and private sector to achieve the Vision 2020 strategic goals. The program focused on capacity creation, effective utilisation of capacity, and capacity retention. See APRM p. 97.

[153] Bahati Morara, 'Enhancement of Corporate Governance Through Privatization' (June 2010) p. 2.

[154] Rwanda Joint Governance Assessment Report p.73.

[155] The African Peer Review Mechanism report noted that stakeholders in Rwanda recognized that establishment of a Capital Market would have the effect of disseminating good corporate governance principles. See APRM report p. 92.

[156]See the Rwanda Stock Exchange website at http://www.rse.rw/site/pages/role+of+rse/5 [accessed 17/11/2016]. Members of the Rwanda Securities Exchange, who provide brokerage services include African Alliance Rwanda Securities, CDH Capital Limited, Faidha Securities Rwanda, Baraka Capital Limited Core Securities Ltd, SBG securities Ltd, BK Securities Limited and Dyer and Blair. Custodians of the stock exchange include BK Custody, and KCB Rwanda SA Custody. Renaissance Capital (Rwanda) Ltd is a sponsoring broker. Credit rating agencies include Metropole Corporation, Bloomfield Investment Corporation, and Agusto & Co. Ltd. Rwanda National Investment Trust Ltd, and Alpha Africa Assets Managers Rwanda Ltd are asset management companies. Aims Capital Limited, MCB Capital Markets, MCB Capital Markets, DKB Markets Ltd, Liaison Financial Services Limited are investment advisors. Companies listed on the Rwanda Stock Exchange include Bank of Kigali, Kenya Commercial Bank (KCB), Equity Bank Group Ltd, Uchumi Supermarket Ltd, National Media Group, Bralirwa, and Crystal Telecom.

[157] Mohamed Al Zuhair, 'Privatization Programs, Ownership Structures, and Market Development: The Role of Country Characteristics in Defining Corporate Governance Standards' (2003) available at http://pqdtopen.proquest.com/doc/193988520.html?FMT=AI [accessed 20/03/ 2017] p. v. Consider Simon Johnson and Andrei Shleifer, Privatization and Corporate Governance (January 2004) p.13-14 available at http://www.nber.org/chapters/c10182.pdf accessed on 20th March 2017, where that author notes that where there is poor corporate governance of privatized institutions, they under perform.

[158] Julian Mwika, 'Is it vital to have Corporate Governance Codes for Institutional Investors & Capital Markets? A Case Study of Rwanda' (28th May 2012) p.37.

[159]Mwika *ibid* quoting Iskander and Chamlou (1998), p. 39. In this respect a key factor that has affected foreign direct investment is Rwanda joining the East African Community in July 2007. As a result of this, Rwanda has

removed many trade barriers, resulting in more foreign companies investing in Rwanda. Businesses from Kenya have invested in Rwanda more than any others from the East African Region, which has contributed to the creation of thousands of jobs. By 2015 Kenya remained the biggest source of Foreign Direct Investment which had by that year invested more than US$450 million. The Kenyan Companies which have invested in Rwanda include Britam, Equity bank, Nakumatt, and Uchumi. See Kenneth Agutamba, 'Rwanda reaping benefits as EAC member – Rugwabiza' (April 13, 2015)

http://www.newtimes.co.rw/section/article/2015-04-13/187793/ [accessed 27/02/2017].

[160] See United States State Department's Office of Investment Affairs' 2015 Investment Climate Statement. Available at https://www.export.gov/article?id=Rwanda-Corporate-Social-Responsibility [accessed 3/03/2017]. The state department also reports that 'in recognition of the firm's strong commitment to CSR, the U.S. Department of State awarded Sorwathe, a U.S.-owned tea producer in Kinihira, Rwanda, the Secretary of State's 2012 Award for Corporate Excellence for Small and Medium Enterprises. In 2015, U.S. firm Gigawatt Global was a finalist for the Secretary of State's Award for Corporate Excellence in the environmental sustainability category'. Concerning corporate social responsibility. See also Richard Barenzi, 'Understanding Corporate Social Responsibility' (28th September 2010) New Times Newspaper (Rwanda) available at http://www.newtimes.co.rw/section/article/2010-09-28/24408/ [accessed 3/03/2017] states that '[b]ig companies like [British American Tobacco] BAT, BRALIRWA, MTN, BCR, [Kenya Commercial Bank] KCB often publicize their Corporate Social Responsibility activities. Yet the concept of CSR is not yet well appreciated by Rwandan businesses or the general public'.

[161] Office of the Ombudsman, 'Rwanda Anti- Corruption Policy' (May 2012) p.6

http://www.minijust.gov.rw/fileadmin/Documents/Policies/Rwanda_Anti_Corruption_Policy.pdf accessed on the 27th March 2017.

[162] *Ibid* p.3.

[163] *Ibid* p.4.

[164]Eugene Kwibuka, 'Ombudsman outlines plans to intensify corruption fight' *The New Times* (Rwanda) (22nd February 2016) available at http://www.newtimes.co.rw/section/article/2016-02-22/197313/ [accessed 27/03/2017].

[165] Ibid

[166] Office of the Ombudsman, *supra* note 57 p.7.

[167] Arthur Levitt, (Speech by SEC Chairman :Remarks Before the Conference on the Rise and Effectiveness of New Corporate Governance Standards) (December 2000)

https://www.sec.gov/news/speech/spch449.htm [accessed 31/03/2017].

Chapter 12

Corporate Governance in Swaziland

Ismail Adelopo and MD Humayun Kabir

12.0 Introduction

Corporate governance is no longer a new term due to recent corporate and international financial scandals which have led to significant appreciation of the importance of corporate governance not only at corporate level but even at a national level (Tsamenyi et al 2007; Gugler et al 2003; Rabelo & Vasconcelos 2002; Reed 2002). This is evidenced in the improved corporate transparency and disclosure practices at corporate and national levels (Bhasin & Reddy 2011; Ho & Wong 2001). Corporate governance continues to be a hotly contested issue in terms of its meaning and boundaries (Naidoo 2002). While corporate governance is presented from a somewhat narrow purview of micro-economic concerns of firms and its constituents by some authors (e.g. Cadbury 1992; Shleifer & Vishny 1997), it has often been presented by international organizations such as the Organization for Economic Co-operation and Development (OECD), International Monetary Fund (IMF) and the United Nations (UN) from much broader and macro-economic perspectives. For instance, Shleifer and Vishny (1997:737) noted that, "Corporate governance deals with the ways in which suppliers of finance to corporation assure themselves of getting a return on their investment". On the other hand, the OECD (2004) suggested that good corporate governance is capable of improving the confidence of domestic investors, lead to reduction in cost of capital and may facilitate access to much needed capital for productive activities. Yet there are others that see corporate governance from a completely different perspective. For example, Bhasa (2004) pointed out that corporate governance concerns the desirability or otherwise of the convergence of global corporate governance best practices and global legal system.

However, over the years, corporate governance literature has largely focused on the issues such as corporate governance policy, business ethics, corporate governance reforms, corporate transparency and disclosure, and corporate financial performance mostly in developed countries. We know little about the factors that influence corporate governance disclosure in developing countries, particularly in less developing African countries. Majority of studies set in the context of developing countries tend to focus more on replicating findings in the developed economies yet there are massive differences in these contexts (Rabelo & Vasconcelos 2002; Mensah 2002). It is therefore very important to investigate the real determinants of corporate governance disclosures in the context of developing economies rather than just the usual replications. While previous corporate governance disclosure studies suggest motives such as capital market transaction, corporate control contest, stock compensation, litigation cost and propriety cost hypotheses for disclosures (Collett & Hrasky 2005; Healy & Palepu 2001), the literature is silent on the motives for disclosures in environments with severely underdeveloped market and political infancy which make identified motives in the literature inconsequential in these unique settings. As a result, this study attempts to provide empirical evidence on the motives for corporate governance disclosure in a less developing country in Southern Africa and within the Commonwealth (i.e. Swaziland).

In 1902, Swaziland became a British colony and was under British control until independence. The country gained full independence in September 1968 and joined the Commonwealth in 1968. As a Commonwealth nation, Swaziland participates in all

Commonwealth activities. Swaziland is a Sub–Saharan African country with poor economic growth. Its governance consists of Swazi traditional and modern parliamentary systems [African Development Bank (ADB) 2005]. The King of Swaziland (King Mswati III) is the head of both systems and has power over legislation and the judiciary (ADB 2005). The country faces serious political and socio-economic challenges. For example, political parties are banned in the country, HIV infection rate is very high (World Development Indicators 2011), unemployment rate stands at 40.6% in 2013 (Swazistats 2015) and poverty level is close to 63% (ADB 2014). Gross Domestic Product (GDP) growth of Swaziland is just 1.7% in 2014 which is far lower than the Sub-Saharan Africa average growth (4.6%) (World Development Indicators, 2014). Further, the country's per capita growth averaged about 0.7% per year between 1994 and 2006 which is also far lower than other Southern African countries such as South Africa (1.6%), Botswana (5.2%) and Namibia (2.1%) (World Bank 2007:2). Between 1990 and 2013, Swaziland's per capita average growth was 0.8% (commonwealth. org). Foreign Direct Investment (FDI) has also fallen drastically in the country, falling from 7.84% of GDP in 2002 to 2.54% of GDP in 2010 (World Development Indicators 2011). According to World Bank's Doing Business Report for 2014, Swaziland was ranked 110th out of 189 countries on ease of doing business (World Bank 2015). Moreover, Swaziland's s civil service wage is the third highest in Africa that accounts for 56% of the country's GDP (IMF 2016) and the lion part of its revenue (more than 60%) comes from Southern African Customs Union (SACU) receipts.

In spite of these miserable social and economic conditions and political infancy in the country, companies in Swaziland particularly public and the few listed companies are required to disclose corporate governance information in their annual reports. Besides, Swaziland Stock Exchange listed companies are required to comply with South African King Report on corporate governance framework where they follow a number of principles and guidelines for corporate governance disclosure practice.

Furthermore, the IMF has recently been invited by Swaziland government for an assessment on the level of corporate governance practice including transparency and disclosure in the country (Adelopo & Kabir 2012). It should be noted that, as a benchmark, Swaziland adopts South African 'King Report on Corporate Governance' guidelines for its own corporate governance practices, as Swaziland does not have any specific corporate governance framework for all its entities.

These are arguably indicative of significant improvements in corporate governance practices in a country that can reasonably be considered obscured and with insignificant incentive for corporate governance reforms. It therefore suggests that there are possibly different incentives for corporate governance practices including disclosure practices than those documented as drivers in the developed economies or even developing economies with considerable incentives for governance reforms and enhanced corporate governance practices in such countries such as South Africa, Nigeria and Egypt. Thus, this paper presents a case to investigate the motives for corporate governance disclosure in a different country context - Swaziland.

The next section provides an overview of corporate governance and disclosure practice in Swaziland including a brief description of the regulatory framework for corporate disclosure in the country. This is followed by a description of the methodology and findings. The last section of the chapter provides the conclusions.

12.1 Corporate governance and disclosure in Swaziland

A large number of African states are registered with the African Peer Review Mechanism (APRM) in order to improve their governance systems and develop a corporate governance code. Swaziland was not registered with APRM until 2013 to take advantage of this mechanism (APRM 2013). However, Swaziland government has undertaken several initiatives such as public sector management reforms and strengthening of anti-corruption measures to improve the corporate governance environment across the public sector and civil service (ADB 2005). The present Swaziland governance system leads to "conflicting public policies" and compromising the principles of good governance since the King of the country is the head of the modern parliamentary system of government and, on the other hand, exercises legislative, administrative and judicial powers in the traditional system (ADB 2005:7-8). "The country's Common law, legislative and judicial activism processes have generally been slow in keeping pace with the legal and regulatory risk and challenges presented by the fast changing domestic and international financial market environment" (Central Bank of Swaziland 2008/2009:41).

The Economic Review and Outlook 2002/03-2006/07 Report prepared by the Ministry of Economic Planning and Development Swaziland (2004:52) stated that "government, as owner of the business, does not fulfil its obligations in terms of controlling the performance of parastatals and assessing the cost-effectiveness of their activities". As noted earlier, Swaziland does not yet have a unique and standard corporate governance framework in order to govern all business enterprises. Private and public sectors including listed companies in the country normally follow a number of principles and guidelines contained in South African 'King III Report' on voluntary basis. However, Ministry of Finance Swaziland (2007) has produced guidelines regarding corporate governance and code of conduct for Public Enterprises (PEs) in a document, namely 'Corporate governance and code of business conduct and ethics for board of directors of public enterprises'. Besides, PEs in Swaziland are governed by a wide range of statutory regulations (Ministry of Finance Swaziland 2007:3). The management of public finances in Swaziland is governed by a number of statutes such as The Finance and Audit Act (1967) as amended in 1992; The Audit Bill (1998), and The Public Enterprises (Control and Monitoring) Act 1989 (ADB 2005:16). The Public Enterprise Unit monitors financial and operational affairs of PEs under a legislative framework provided by the Public Enterprise (Control & Monitoring) Act 1989 (ADB 2005:12). The regulations with regard to accounting records of listed companies are set out in accordance with International Accounting Standards along with the new Companies Act 2009 and Swaziland Stock Exchange requirements. It should be noted that Swaziland's business organizations were governed by the old Companies Act 1912 for almost a century. However, the Swaziland Stock Exchange, the Swaziland Institute of Accountants, civil society organizations, Swaziland Central Bank, and Ministry of Finance have become very vocal in calling for more visible corporate governance changes in the country.

12.1.1 Legal framework for corporate disclosure in Swaziland

Legal and regulatory frameworks are country specific, and impact on disclosure practices (Akhtaruddin 2005). In Swaziland, accounting and corporate disclosure are underpinned by requirements specified in the Companies Act 2009 and Swaziland Stock Exchange. For instance, listed companies are also required to comply with the specific requirements concerning corporate governance and disclosure. Listed companies are required to disclose relevant information in the annual financial statements in compliance with Swaziland Stock Exchange requirements such as borrowings, headline earnings per share, directors' interests in the share capital, profit forecasts, directors' fees, salary, bonus, share options, etc. The Companies Act 2009 also addresses issues related to accounting records and disclosures. For example, in accordance with Section 245 of the new Act, "every company shall keep such accounting records as are necessary fairly to present the state of affairs and business of the company and to explain the transactions and financial position of the trade or business of the company" (Swaziland Companies Act 2009:110). Financial statements are prepared in accordance with International Financial Reporting Standards and the requirements of Swaziland Companies Act 2009.

The Swaziland Institute of Accountants is the only professional accounting body in Swaziland. Since Swaziland Institute of Accountants is not a member of the International Accounting Standards Board (IASB), the institute is not legislatively strong in terms of issuing accounting and reporting standards. Consequently, the institution persuades the government for the adoption and implementation of standards issued by the IASB. Swaziland Institute of Accountants fosters acceptance of International Accounting Standards and International Standards on Auditing. However, besides the International Accounting Standards, Swaziland complies with the disclosure provisions specified in the Companies Act 2009 which are mandatory (see, Section 256 - 258 of Swaziland Companies Act 2009).

12.1.2 Corporate governance disclosure practice in Swaziland

The corporate governance principles in respect of financial transparency and information disclosure practice have become more prominent around the world especially due to the global financial crises. Bashin and Reddy (2011:393) suggested, "corporate governance disclosure is a fundamental theme of the modern corporate regulatory system". According to OECD (2004), the corporate governance framework should ensure that all material matters regarding financial situation, performance, ownership and governance of the company are disclosed accurately and timely. Adequate corporate "disclosure helps improve public understanding of the structure and activities of companies, their policies and performance" (Hong Kong Society of Accountants 2001:4) and subsequently, it improves transparency and results in good corporate governance practice (Bashin & Reddy 2011).

In Swaziland, as in many African countries, governance transparency, accountability and disclosure issues have not been addressed seriously due to high incidence of political favouritism in the corporate, and public sector in particular (Dlamini 2005). Although Swaziland's business sector is mainly dominated by small and family owned enterprises, the problems and challenges facing Swaziland government including public sector in terms of poor governance constitute a demand for effective corporate governance practices, transparency and honest disclosure. Therefore, an analysis of the motives for corporate governance disclosure in the context of Swaziland as a less developing African country will be of great value to academics and business managers to understand the disclosure practice mechanisms in a setting with poor market infrastructure and poor economic growth.

12.2 Methods

The study relies on documentary evidence, conceptual analysis and personal observations of one of the researchers who lived and worked in the country. The personal observations mainly consist of one of the researchers' work experience in our focal country for a decade as an academic in one of the universities that provide unique insights on the external motives for corporate governance disclosure in the country. Previous studies have used observations and informed observer approach as the main method for research (e.g. Gobind & Ukpere, 2012; Uddin & Choudhury, 2008). This method is consistent with the tenets of qualitative research approach where reality is socially constructed and interpreted through the prism of the social actor. A fundamental requirement for authentication is that such social actor should possess authoritative knowledge and ability. Further, in order to investigate the motives for corporate governance disclosure, the study adopts the framework inherent in the idea of Di Maggio and Powell's (1983) institutional isomorphic. The implication of the deficiency in the majority of the previous studies and their prescriptions on corporate governance motives, mechanisms, and outcomes in a context such as Swaziland requires reconsideration in the context of the external environmental influence on firms' governance paraphernalia. This study therefore uses Di Maggio and Powell's (1983) three types of isomorphism forces (coercive, mimetic, and normative) to argue and understand the external drivers for corporate governance disclosure in Swaziland. Di Maggio and Powell's three isomorphic forces are used in identifying and determining the external pressures for corporate governance disclosure practices in jurisdictions with no incentive for corporate disclosure but which nonetheless requires adherence to corporate governance disclosure practices suitable for jurisdictions with sufficient incentive for such disclosure. This paper is also based on the review of documents of both national and international organizations such as the World Bank, IMF, UNO amongst others.

12.3 Findings and Discussion

The analysis in this section is undertaken using the institutional isomorphism lenses and personal and longitudinal observations.

12.3.1 The external coercive drivers for corporate governance disclosure in Swaziland

This study identified two main sources of the coercive forces for corporate governance disclosure in the country. First, the pressure from international organizations and secondly, pressure from regional arrangements with other neighbouring countries. According to Di Maggio and Powell (1983), coercive pressure could be legislative or persuasive; and could be multi-layered, linked to benefits by the focal organization from a conferring organization. They all aim, inadvertently, to drive the focal actors towards actions intended to maintain legitimacy with its conferring publics. In terms of the international organizations, the study identified the influential roles of the World Bank Group, IMF and the OECD's corporate governance principles while the regional influences appear to come mainly from the Southern African Development Community (SADC) and SACU. We now take these in turn.

12.3.2 World Bank Group, IMF and the OECD Corporate governance principles

In 1999, both World Bank (WB) and IMF started the standard and code initiative, now popularly referred to as the Reports on the Observance of Standards and Codes (ROSC). This initiative enables the organizations to assess member states' compliance and observance of international standards and codes in 12 key areas including corporate governance. Countries' corporate governance postures are benchmarked against the OECD principles. Based on the overall corporate governance situation in Swaziland, ADB (2005:27) stated, "Swaziland government has a long way to go in complying with the OECD Principles of Corporate Governance". In complying with the OECD Principles of Corporate Governance, ADB (2005:46) recommended that it is necessary for Swaziland government to review the existing regulatory framework and corporate governance systems including disclosure practices.

Although assessment is at the request of a member state, however, non-compliance or poor report could have detrimental implications for the country concerned, especially developing countries. This is coupled with the well-documented influence both organizations have in determining the internal policy direction of client countries (Dreher 2006; Taylor 1997). Poor showing in the ROSC could have adverse effects on the investment destination of global investors. Apart from this, to benefit from the WB and IMF loans, countries would have to conform to required levels of economic and governance transparency and accountability. The initiative has assessed and reported on over 100 countries so far. It is therefore becoming a vital tool for assessing the state of corporate governance at national level and invariably pushing countries towards common corporate governance practice. It has been observed that international organizations dominate the countries' choice of corporate governance development (Soederberg 2003). Thus, it is not surprising that the Swaziland government has eventually invited IMF for an assessment of compliance with codes and standards in the country through the ROSC initiative. This is an example of a subtle or persuasive external pressure towards corporate governance disclosure practice in Swaziland.

12.3.3 Regional arrangements (SADC and SACU)

Swaziland has bilateral arrangements with countries in the Southern African region including through the Southern African Development Community (SADC) and Southern African Customs Union (SACU). The SADC emerged from the Frontline State; an organization comprising mainly Southern African states set up to facilitate attainment of independence by member states. However, following full independence by most member states, its focus has now moved towards good governance and economic development within the region. The SADC through its objectives that see a common future and path for the countries in the region, and institutions such as the Development Finance Resource Centre organizes capacity building and training programmes for member states on several issues including corporate governance.

Furthermore, the SACU also provides another strong platform for common economic and governance pattern in the region through common custom arrangement. Revenue from SACU is deposited in a common revenue pool, which is under the custodian of South Africa. Botswana, Lesotho, Namibia, and Swaziland (BLNS) countries get nearly half of the SACU revenue though the total GDP of BLNS countries is less than 10% of SACU member states' GDP (www.sacu.org). Swaziland relies heavily on SACU receipts for years; more than 60% of Swaziland's total revenue comes from SACU receipts. South Africa as the major economy in the region makes the most contribution to the fund and consequently wields sufficient influence in its distribution.

These regional relationships require compliance with common pacts and practices. With South Africa as the leading economy in the Southern African region, most of the corporate governance agenda in the region as expected, are generated from the country. It is therefore not surprising that Swaziland and other countries in the region are following South African footsteps in their corporate governance ideas including disclosure practices. We argue that both international organizations and bilateral cooperation provide coercive pressure for Swaziland to implement and adhere to corporate governance and disclosure requirements even with minimal incentive for this in the economy.

12.3.4 The external mimetic drivers for corporate governance disclosure practice in Swaziland

Di Maggio and Powell (1983) described mimetic pressure as threat to legitimacy arising from strategic ambiguity and uncertainty, which may lead focal organizations to adopt imitation and modeling of organizations considered more successful as solution to perceived legitimacy gaps.

Swaziland is experiencing a period of sustained poor economic performance evidenced by falling FDI, especially since the end of Apartheid in South Africa in 1994, and high reliance on revenue from SACU. On the other hand, South Africa maintains an impressive economic performance, as the largest and most developed economy in Africa. It is the largest trading partner with Swaziland with about 70% of its exports going into South Africa (World Development Indicators 2011) and is also its main source of imports.

The country appears to be modeling its policies on South Africa in order to improve its own economic fortunes. For example, its currency is pegged at 1:1 with South Africa's Rand, which requires substantial coordination by the South African authority and indicates significant dependence on the monetary and fiscal policy of its superior neighbour. There is also a heavy presence of South African businesses and expertise in the country's economy especially in the banking and service sectors. Perhaps a more obvious evidence of imitation and modeling is in respect of the corporate governance reforms and code. South Africa has one of the enviable corporate governance reforms and King I, II and III are one of the most popular reports on corporate governance. It is therefore not surprising that Swaziland adopts the South African King report for corporate governance practice in the country. This shows the mimetic influence that gradually leads to homogeneity in social actors' strategy, structure and outcomes.

Furthermore, according to Di Maggio and Powell (1983), sharing common resource base could facilitate mimetic pressure. For example, common training provider or consultancy could lead to knowledge transfer with which the focal organization may reduce their perceived legitimacy gaps. As mentioned earlier, the SADC through its organs and institutions facilitates cooperation and coordination among member states on various issues including corporate governance, which could eventually lead to common features and outcomes amongst member states.

12.3.5 The external normative drivers for corporate governance disclosure practice in Swaziland

Normative pressure can be by the professional organizations through providing training and knowledge (Di Maggio & Powell 1983). For instance, the study identifies that Swaziland Institute of Accountants, which is affiliated with various regional and international professional organizations (such as International Federation of Accountants and Eastern, Central and Southern African Federation of Accountants) provides regular training on corporate governance issues. Another example is the Corporate Skills Centre of Swaziland funded by all the banks in Swaziland, which has recently signed a cooperative agreement with a South African professional institution (i.e. Milpark Business School) for training and skills improvement for the professionals working in various sectors in the country (Nkambule 2011).

Furthermore, for public sector employees' skill development and training, Swaziland government normally hires consultant firms from abroad, mainly from South Africa due to lack of skill training centres in the country. The top management of accounting and auditing firms in Swaziland is mainly dominated by South African Chartered Accountants. Since Swaziland does not have professional accounting and auditing examination bodies, most of the accounting and auditing professionals working in the country are qualified from international and regional professional examination bodies such as the Association of Chartered Certified Accountants, South African Institute of Chartered Accountants and Institute of Internal Auditors South Africa. It suggests that global awareness on corporate governance is easily transferred into the country by the country's professional affiliation with international associations and the professionals with foreign qualifications practicing in the country. Consequently, it implies that these are the sources of normative pressure for corporate governance practice and disclosure in Swaziland.

12.4 Conclusion

The chapter aims to understand the external drivers for corporate governance disclosure in a less developing country within the Commonwealth. Consequently, this study considered corporate governance disclosure drivers in a unique market setting characterized by poor market infrastructure and poor economic growth. The study finds that drivers for corporate governance disclosure practices in a developing country (e.g. Swaziland) are mainly derived from external institutional framework because of less pressure from internal arrangements. Furthermore, the study finds that pressure from international organizations, bilateral and regional relationships, affiliation of professional associations to international and regional accounting bodies and global development in corporate governance with its effects on professional training and syllabus contents are also key drivers for corporate governance disclosure in a less developing country. Moreover, the less developing countries would have to implement economic reforms and conform to required levels of governance transparency and accountability in order to qualify for loans and other social and economical support from external sources since the capital markets and markets for corporate control are virtually non-existent in less developing countries.

This study articulates the drivers for corporate governance disclosure in Swaziland with poorly functioning markets and thereby improves our understanding on their potential policy and regulatory implications. The study shows the importance of considering a mid-range theoretical paradigm for the discourse of corporate governance. This is more so the case when due consideration is given to the institutional context that determines governance mechanisms especially in jurisdictions with severe structural and infrastructural imbalance. This study is motivated due to the dearth of studies on corporate governance practices in the developing economies and the dare need for greater understanding of how this phenomenon is construed in such context (Dahawy 2009), together with the increasing interest in corporate governance in Swaziland as a developing African economy.

REFERENCES

Adelopo I.; Kabir Md H. 2012. Corporate Governance Reforms: Between Instrumentality and Institutional Drivers, *Corporate Ownership and Control* 10(1): 659-674.

African Development Bank 2005. Kingdom of Swaziland – *Country Governance Profile*. Operations Department, ONCB, ADB.

African Development Bank 2014. Swaziland Economic Outlook. *African Economic Outlook 2014*, ADB.

African Peer Review Mechanism 2013. *Africa's Self Assessment for improved good governance. Member Countries, APRM.* Available at: http://aprm-au.org/aprm-map. (Accessed on 24 March, 2013).

Akhtaruddin M. 2005. Corporate mandatory disclosure practices in Bangladesh, *The International Journal of Accounting* 40(4): 399-422.

Bhasa MP. 2004. Global corporate governance: debates and challenges, *Corporate Governance* 4(2): 5-17.

Bhasin M L.; Reddy, M. 2011. Corporate Governance Disclosure Practices: The Portrait of a Developing Country, *International Review of Business Research Papers* 7(1): 393 – 419.

Cadbury Sir Adrian 1992. *Report of the Committee on the Financial Aspects of Corporate Governance.* Gee & Co. Ltd., London.

Central Bank of Swaziland 2008/2009. *Annual Report 2008/2009.* Central Bank of Swaziland, Swaziland.

Collett P; Hrasky S. 2005. Voluntary Disclosure of Corporate Governance Practices by Listed Australian Companies, *Corporate Governance: An International Review* 13(2): 188-196.

Commonwealth.org. Swaziland – *Swaziland History.* Available at: http://thecommonwealth.org/our-member-countries/swaziland. (Accessed on 22 February, 2018).

Dahawy K. 2009. *Developing Nations and Corporate Governance: The Story of Egypt.* Global Corporate Governance Forum, The International Financial Corporation. Available at: http://www.ifc.org/ifcext/cgf.nsf/AttachmentsByTitle/PaperKhaledDahawy/$FILE/DahawyKahled4.pdf. (Accessed on 30 April, 2011).

Di Maggio PJ.; Powell WW. 1983. The iron cage revisited: Institutional isomorphism and collective rationality in organizational fields, *American Sociological Review* 48: 146-160.

Dlamini MP. 2005. Public Enterprises and the Challenges of Privatization in Swaziland, *African Journal of Public Administration and Management* XVI(2): 29-49.

Gobind J; Ukpere WI. 2012. Accessing HIV awareness and prevention information: South African challenges, *African Journal of Business Management* 6(50): 11999-12004.

Dreher A. 2006. Does Globalization Affect Growth? Evidence from a new Index of Globalization, *Applied Economics* 38(10): 1091-1110.

Gugler K; Mueller D; Burcin Y. 2003. The Impact of Corporate Governance on Investment Returns in Developed and Developing Nations, *Economic Journal* 113(491): 511-539.

Healy PH; Palepu KG. 2001. Information Asymmetry, Corporate Disclosure, and the Capital Markets: A Review of the Empirical Disclosure Literature, *Journal of Accounting and Economics* 31(1-3): 405-440.

Ho SSM; Wong KS. 2001. A study of the relationship between corporate governance structures and the extent of voluntary disclosure, *Journal of International Accounting, Auditing & Taxation* 10:139–156.

Hong Kong Society of Accountants 2001. Corporate Governance Disclosure in Annual Reports: A Guide to Current Requirements and Recommendations for Enhancement. *Hong Kong Society of Accountants, March 2001*: Hong Kong.

International Monetary Fund 2016. *Case Study on Management of Government Compensation and Employment Institutions, Policies & Reform Challenges* Available at: www.imf.org/external/np/pp/eng/2016/040816. (Accessed on 11 January, 2017)

Mensah S. 2002. *Corporate Governance in Ghana: Issues and Challenges*. Paper presented at the African Capital Markets Conference, December 2002.

Ministry of Economic Planning and Development Swaziland 2004. *Economic Review and Outlook-2002/03-2006/07*. Ministry of Economic Planning and Development, Swaziland Government, Swaziland, (March 2004 Report).

Ministry of Finance Swaziland 2007. *Corporate governance and code of business conduct and ethics for board of directors of public enterprises.* Public Enterprise Units, Ministry of Finance, Swaziland Government, Swaziland.

Naidoo R. 2002. *Introduction to Corporate Governance.* In Naidoo, R. (Ed.). Corporate Governance: An Essential Guide for South Africa Companies, Double Story Books: Cape Town.

Nkambule N. 2011. *Banks must be smart, tight in managing risks.* The Swazi Observer, Mbabane, 7 July 2011. Available at: http://www.observer.org.sz/index. (Accessed on 17 July, 2012).

Organization for Economic Co-operation and Development 2004. *Principles of Corporate Governance.* Available at: http://www.oecd.org/daf/corporate/principles. (Accessed on 25 December, 2005).

Rabelo FM; Vasconcelos FC. 2002. Corporate Governance in Brazil, *Journal of Business Ethics* 37(3): 321-335.

Reed D. 2002. Corporate Governance Reforms in Developing Countries, *Journal of Business Ethics* 37(3): 223-247.

Shleifer A; Vishny RW. 1997. A Survey of Corporate Governance, *The Journal of Finance* 52(2): 737-781.

Soederberg S. 2003. The Promotion of 'Anglo-American' Corporate Governance in the South: Who Benefits from the New International Standard? *Third World Quarterly* 24(1): 7-27.

Swaziland Companies Act 2009. *The Companies Act Notice 2009.* Swaziland Government, 25 December 2009, Mbabane, Swaziland. (Government Gazette Vol. XLV).

Swazistats 2015. *Swaziland Unemployment.* Kingdom of Swaziland Statistical Office. Available at: http://www.swazistats.org.sz. (Accessed on 17 November 2015).

Taylor L. 1997. The Revival of the Liberal Creed - the IMF and the World Bank in a Globalized Economy, *World Development* 25(2): 145-152.

Tsamenyi M; Enninful-Adu E; Onumah J. 2007. Disclosure and corporate governance in developing countries: evidence from Ghana, *Managerial Auditing Journal* 22(3): 319-334.

Uddin S; Choudhury J. 2008. Rationality, traditionalism and the state of corporate governance mechanisms: Illustrations from a less-developed country, *Accounting, Auditing & accountability Journal* 21(7): 1026-1051.

World Bank 2007. Swaziland – An Assessment of the Investment Climate. Financial and Private Sector Development Africa Region, *The World Bank, Washington, DC.* (Volume II, Report No. 43637-SZ).

World Bank 2015. *Doing Business 2015 – Going Beyond Efficiency, Economic Profile 2015: Swaziland.* The World Bank, Washington, DC.

World Development Indicators 2011. *Swaziland – Country Brief.* World Bank. Available at: http://data.worldbank.org/data-catalog/world-development-indicators/wdi-2011. (Accessed on 20 April, 2012).

World Development Indicators 2014. *Swaziland – Data.* World Bank. Available at:

http://data.worldbank.org/country/swaziland. (Accessed on 17 November, 2015).

Www.sacu.org. Southern African Customs Union. South Africa. Available at: http://www.sacu.org.

Chapter 13

Corporate Governance in Tanzania

Samuel E. Fulgence

13.0 Introduction

The collapse of major corporations in the UK and the USA drew the world's attention to corporate governance, which has now become a global phenomenon. The Asian economic crisis also contributed to raising the profile of corporate governance as the crisis has been linked to poor corporate governance practices (Fulgence, 2013). The notion of shareholder value, promoted by the conservative governments in the UK and the US in the early 1980s, had a significant impact on corporate governance developments. The corporate scandals in Tanzania, such as EPA, MEREMETA, DOWANS and RICHMOND which occurred between 2000 and 2008 are believed to originate from the lack of accurate reporting, transparency and negligence of adequate procedures (Fulgence, 2103). These scandals negatively impacted the national economy, and affected the majority of shareholders, stakeholders, practitioners, politicians and academics (Fulgence, 2013).

The scandals alarmed the Government, causing it to strengthen its financial system and improve corporate governance. The government was forced to establish the Public Procurement Regulatory Authority (PPRA), which make compulsory separation of the duties of accounting officers and procurement officers. It also imposed updated procedures on tendering and consignment contracts (Fulgence 2013).

There have been attempts recently to address the challenge of corporate governance in Tanzania. The privatization policy, the establishment of the Banking and Financial Institutions Act in 1991, and its Regulations in 1997 are significant developments in this regard. In 2002, the Capital Markets and Securities Authority (CMSA) and the Steering Committee on Corporate Governance in Tanzania introduced separate but related sets of principles for effective corporate governance, which companies are being encouraged to adopt and implement (Melyoki 2005). The Institute of Directors Tanzania (IoDT) was established May 2012 so as to promote the compliance with corporate governance principles. Amongst its objectives is the development of codes of corporate governance for private sector institutions.

This chapter explores the historical development of corporate governance in Tanzania, drawing attention to the internal and external factors that have aided this development, and examines the nature of corporate governance guidelines and practices introduced in Tanzania and its current status.

13.1 Brief Background of Tanzania

Tanzania is a developing country, which is located in the East African region. It is positioned around 5° south of the Equator and 45° east and covers 945,087 sq. km. It has a border with Uganda and Kenya in the North and North East respectively, and in the North West and West by Rwanda, Burundi and the Democratic Republic of Congo. To the South and Southwest lie Mozambique and Zambia respectively, and the Indian Ocean lies to the East. Tanzania is made up of Tanganyika and Zanzibar.

The country operates under a multiparty political system. It was a single-party state from 1965 to 1991. In 1992 a multiparty political system was re-introduced. Since 1992 Tanzania has been governed as a multiparty democracy, and it is currently in the process of adopting a new constitution. The multiparty political system was re-established following the recommendations of the Nyalali Commission despite the fact that this Commission found 77.2% of the people interviewed were for the single party system.

Tanzania is one of the most politically stable countries in Africa since independence[168]. The United Republic of Tanzania is the union of two sovereign states namely Tanganyika, and Zanzibar in 1964. Tanganyika regained its independence in 1961 from the British and became a Republic in 1962. Zanzibar regained its independence from the Arab Sultanate in 1964. The two joined to form the current union – called the United Republic of Tanzania (URT) - in April 1964. The Act of Union spells out the functions that are union matters and those that are the responsibility of each member. Also, the Act of the Union recognizes that most governance issues are non-union matters, and to the extent possible, attempts to include a discussion of the issues as they relate to Tanzania mainland and the island of Zanzibar.

Tanzania is a low-income developing country in East Africa. It is currently in a period of transition since adopting an economic liberalization strategy in the mid-1980s. This included embarking on a structural adjustment program supported by the World Bank and the IMF aimed at transforming its economic infrastructure from a socialist-oriented, state-ownership economy to a market-oriented, private sector-led economy.

The country is highly aid dependent, with around 14% of its GDP derived from official development assistance through various programs with a range of development aid partners. The economy has to date been largely agrarian, with the agricultural sector contributing 27 to 32 percent of its GDP in the period 2007 to 2013. The sector provides employment to around 75% Tanzania workers. The growth of the agricultural sector has been lower than the growth of the overall economy, constrained by infrastructure gaps and low productivity.

Under Tanzania's Development Vision 2025, the national strategy is to encourage social and economic development. The country aims to transform into a middle-income economy by 2025. Adoption of a market approach in which the private sector is the engine of economic growth is central to the national strategy (Van, 2013). In line with this, the government is currently undertaking a roadmap of sectoral and macro reforms including some aimed at improving the investment centre. Under these reforms, Tanzania has experienced strong economic growth since the 1990s. While Tanzania's overall macroeconomic performance remains strong, with economic growth of more than 7 percent (p.a) achieved in 2013 and 2014, and forecast economic growth broadly in line, this performance has not engendered broad-based, inclusive development or structural transformation. Despite this increase, Tanzania continues to experience high poverty levels by regional and international standards, with an average of per capita income significantly lower than that of the Sub-Saharan Africa average and around 30 per cent of its population living below the poverty line. In light of this, Tanzania's current infrastructure transformation and development aim to achieve more broad-based results to address poverty reduction.

Decades of public sector reforms have enabled Tanzania to perform better than most African countries on public finance management (APRM, 2013). Despite its satisfactory progress with reforms there are mixed results in the areas of governance and corruption. While corruption is pervasive in all aspects of political and commercial life, it is considered particularly prevalent in the energy and natural resources sectors (Linda, 2014).

More recently the significant drivers of economic growth have been information and communication, construction and manufacturing. The most important transformative factor on the economy, however, is the vast natural gas reserves discovered in 2012-2013 through oil and gas exploration activities, which are expected to draw substantial private capital and a net inflow of FDI in Tanzania (Business Times, 2014).

Economic performance has remained stable and steady over the past years. The results show that there was 7 per cent growth in 2014, and preliminary estimates indicate the same growth rate in 2015[169], driven primarily by the services, industry, construction, and information as well as communication sectors. For the medium term, growth is projected to outperform the records of 2014 and 2015, increasing to 7.2 per cent[170] in 2016. While other industries are expected to at least perform at their recent levels, the higher growth performance is expected significantly from increased industrial activities and investment in infrastructure. Favourable weather conditions and the reduction in the rate of inflation between 2014 and 2016[171] led to a sustained level of agricultural output and prudent fiscal and monetary policy management. The government's total debt is sustainable at 28.6 per cent, 30.6 per cent and 32.4 per cent of GDP in 2015; 2016 and 2017 respectively (IMF, 2016).

13.2 Evolution of Corporate Governance in Tanzania

Before Tanzania's economic liberalization program, the economy was dominated by the state, and the private sector played an insightful role. State-owned corporations were the dominant form of organization in the country's socialist era, and consequently, Tanzania's industry is underdeveloped in comparison with those of its neighbouring countries (Mo Ibrahim Foundation, 2015). Accordingly, Tanzania now faces many challenges in promoting the development of its private sector, which is dominated by small enterprises. Linked to this, the concept of corporate governance is very much in its infancy in Tanzania, and has to date received little attention (Fulgence 2013; 2014). There is little debate on corporate governance issues reflecting the underlying lack of awareness of its importance for economic development (ACGN 2016:81).

Tanzania being relatively new and small, its growing private sector is being sensitized to the need to implement sound corporate governance standards and practices, especially with support from international financial institutions such as the World Bank and IMF (ACGN 2016:82). In common with the situation that often prevails in developing countries, private-sector entities in Tanzania tend to be characterized by the concentration of shareholder ownership. The Companies Act focuses on the need to maintain control given the perceived weaknesses surrounding capacity to enforce company legislation aimed at protecting shareholder rights. The current corporate governance practices are influenced by the institutional and behavioural legacies of the previous state-dominated economy (ACGN 2016:83). Market participants shield the state-owned corporations from the disciplines of control and accountability. In these corporations, corruption, managerial incompetence, political interference and government subsidization of falling enterprises were reportedly prevalent (Sitta 2005). It is the lack of accountability and effective control in these corporations that were among the factors contributing to their poor performance and failure (Melyoki 2005).

Entities in the banking and insurance sectors are already well-oriented in implementing corporate governance standards through the system of prudential regulation and supervision. Similarly, the listed companies are required to implement the corporate governance guidelines applicable to listed companies. Within the framework of Tanzania's Vision 2025 strategy, institutions are evolving that are focused on advancing the interests of private sector entities. For instance, The Tanzania National Business Council (TNBC) is a forum for public-private sector dialogue on the main issues related to Tanzania's social and economic development. Other industry associations and bodies in Tanzania that facilitate dialogue on strategic matters related to business environment and with government are The Confederation of Tanzania Industries, The Tanzania Private Sector Foundation and The Tanzania Chamber of Commerce and Industry Association (ACGN 2016).

As a relatively new institution, The Institute of Directors Tanzania (IoDT) has yet to follow the example of other African Institutes of Directors in producing a code of corporate governance to ensure best practices for private sector entities in Tanzania. However, initial work on developing a system is underway.

13.2.1 Corporate Governance in the Context of Tanzania

The central planning system for economic coordination in Tanzania and the ownership of corporations by the state implies that the significant experiences of corporate governance will be related to state-owned companies. Between 1967 and 1995, state-owned companies were the most common type of large corporations found in Tanzania. In these companies, corruption, (embezzlement and nepotism) managerial incompetence, political interference and government subsidization of failing corporations were the dominant characteristics of corporate governance. Control and accountability became the prime casualty within these companies. Melyoki (2005) points out that the lack of responsibility and effective control of these corporations left the managers with unfettered powers. They attribute these problems to the ambiguous property rights in the state-owned companies.

Melyoki (2005) reports the paucity of corporate governance in sub-Saharan Africa arising from the ambiguous relationship between the state, as the owner of the corporations, the boards of directors and senior management. The poor governance of state-owned firms in Tanzania has resulted in dismal performance and the failure of these corporations (Melyoki 2005; Fulgence 2013). One result of the poor performance of these organizations has been their inability to provide the necessary "push" for the attainment of social and economic development as envisaged by the post-independence government (United Republic of Tanzania (URT) 1999).

Events in the African region and worldwide have also created the need to develop an understanding of corporate governance practices in Tanzania. Tanzania being a member of the African Union (AU) is required to join the comity of other African countries in improving corporate governance practices. The Africa Union has established a development vision in which member states are individually required to implement initiatives to improve corporate governance practices within them. This concept, called New Partnership and Development [NEPAD], recognizes corporate governance as one of the key issues to be addressed so as to achieve social and economic development on the African continent. Corporate governance plays a pivotal role in poverty alleviation since it improves efficiency in the allocation of resources (Melyoki 2005:13).

Interest in Corporate Governance issues within the Commonwealth countries had its origin in performance improvement efforts and state enterprises privatization. Beyond the African region, Tanzania being a member country of the British Commonwealth, has agreed to undertake measures to improve corporate governance practices. The Commonwealth Association for Corporate Governance (CACG) points out that corporate governance is important in improving the competitiveness of member states in attracting capital and enhances the performance of corporations.

The launch of the IoDT in May 2012 is evidence that effective governance is fundamental for any country's economy regardless of the level of economic development. IoDT has initiated training in Certificate of Directorship (CiDir) program for about 100 directors and senior executives from various organizations in Tanzania to ensure that they get better management expertise to improve the business environment.

Corporate governance in Tanzania is essential to contribute to the development of the private sector, which is currently viewed as the "engine of economic growth." Development of the public and private sectors depends, among other things, on the steady supply of long-term capital for which the effectiveness of corporate governance is a critical factor (OECD 2014; World Bank 2000; Melyoki 2005; Fulgence 2013; 2014). Developing countries, such as Tanzania, face the challenge of attracting the foreign capital flow needed to stimulate the economy. Improving corporate governance contributes to improving a country's competitiveness in attracting foreign capital flow. **Figure 1.1** shows that foreign direct investment (FDI) flows to Tanzania have been increasing over the last ten years, but that there have also been significant fluctuations. The high levels of FDI in 2013, 2014[172] and 2016[173] reflect substantial investments inflows in the mining sector (URT, 2005, 2012). The corporate governance issues are supposed to be maintained and encouraged so as to facilitate further investment inflows into the economy.

Fig 1.1 FDI Inflows to Tanzania from 2002 to 2016

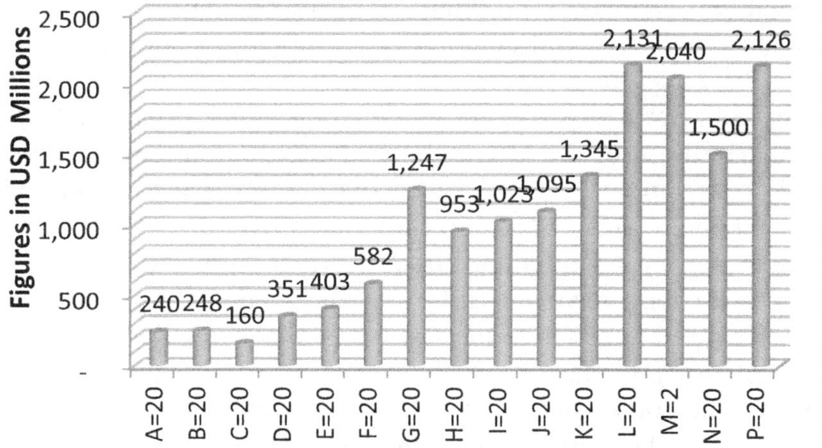

Source UNCTAD World Investment Report 2012;

http://www.tanzaniainvest.com/fdi **and** http://www.tic.co.tz/menu/317

13.2.2 Legislations Governing Corporate Governance in Tanzania

The Tanzanian legal system is based on English common law, statutes and case law. Economic reforms included attempts to evolve local resources of capital for development of private-sector business, and the Tanzanian government has enacted new or revised legislation to establish the framework of laws and regulations that comprise the corporate governance framework for Tanzanian companies (ACGN 2016:82).

The main regulatory framework for corporate governance in Tanzania is the 1992 Public Corporations Act, the 1994 Capital Markets, and Securities Act, and the 2002 Companies Act, which came into force on 1st March 2006. These guidelines were developed, taking into account several other jurisdictions including the United Kingdom, Malaysia, South Africa, and the Commonwealth Association for Corporate Governance as well as OECD Principles of Corporate Governance. Therefore the Capital Markets and Securities Act are modeled on international standards, particularly those of the UK and are also based on the '*comply or explain*' principle. This UK Corporate Governance Code was published on 28th May 2010 and came into effect on 29th June 2010. Although merely a guideline, it requires that companies comply or explain why they have not followed the Code. These guidelines were developed to promote the standards of self-regulation to bring the level of governance in line with international standards. However, in 2009, the World Bank reported that Tanzania was below the OECD average. Since the very recent advent of the East African Common Market, the denationalization of the Securities Market has not taken effect yet. The capital account is yet to be fully liberalized in Tanzania to enable citizens to participate in the cross listing.

The enactment of Tanzania Capital Market and Securities Act 1994 established the Capital Markets and Securities Authority (CMSA) to regulate the capital market in Tanzania and promote Tanzanian securities markets and create the stock exchange. The Capital

Markets and Securities Act contain a list of recommended best practices in corporate governance. The establishment of the Dar Es Salaam Stock Exchange (DSE) with effect from 1997 is aimed at regulating the activities of listed companies through its listing requirements. The businesses seeking a listing on the exchange are required to comply with the Corporate Governance practices by listed companies in Tanzania (CMSA Guidelines) issued by CMSA in 2002.

The guideline further requires the responsibilities of the board of directors to be defined, and outline the appointment and qualifications of an active board and the remuneration of the directors. All listed companies are required to ensure equitable treatment of shareholders, including the minority shareholders. All shareholders should get relevant information on the company's performance through distribution of regular annual reports and accounts. Every shareholder is supposed to receive information on voting rules and procedures, to participate and vote at the general shareholders meeting. Also, shareholders should be entitled to ask questions or seek clarification on the company's performance (Melyoki 2005; Fulgence 2013).

The Companies Act 2002 came into effect on 1st March 2006. The Act introduced a modern shareholder-oriented legal framework for corporate entities along the lines of the Anglo-Saxon model of corporate governance. The Act replaced previous out-dated company legislation. The Business Registration and Licensing Agency (BRELA) administer the Companies Act in Mainland Tanzania and the Registrar General in Zanzibar.

The enactment of Bank of Tanzania Act No. 5 of 2006 and Banking and Financial Institutions Regulations 2013; 2014; 2015 and many others[174] form a cornerstone towards supervision of banking and other financial institutions. The primary objectives of these Regulations are (i) to ensure that banks and financial institutions maintain a level of transparency that is necessary to enable depositors, creditors, and the public to make informed decisions; (ii) to ensure that financial institutions and banks have adequate credit and investment policies to identify and manage business risks;[175] (iii) to enforce compliance with governance issues and other financial reporting standards within the jurisdiction and globally.

13.3 Corporate Governance Across Various Sectors in Tanzania

Most businesses in Tanzania are non-listed, private family owned businesses and owner-managed enterprises where the shareholders and more often the directors are the same people. These reasons and other factors might have contributed to the current embryonic status of corporate governance.

The implementation and compliance approach for corporate governance in Tanzania varies from the nature of the entity, the business activities undertaken and legal form and ownership structure. The remaining sections review corporate governance practices and compliance strategies of (i) banking and financial institutions; (ii) public sector institutions; (iii) Private sector institutions; and (iv) Higher Learning institutions.

13.3.1 Corporate Governance Practice in Banking and Financial Institutions

The Bank of Tanzania has the duty to control and enforce the laws applicable for the prudential supervision of banks and financial institutions. The guidelines for Board of Directors of Banks and Financial Institutions 2008, issued under the Banking and Financial Institutions Act 2006 apply to directors of all banking institutions in Tanzania.

The proper conduct of a bank or financial institution requires that the board of directors function appropriately and at high standards. The responsibility for the oversight and direction of banking institutions rests with the Board of Directors. In discharging his duties effectively, a director must command a high level of integrity, honesty, competence and ability to adhere to effective corporate governance principles (Fulgence 2013). From a banking industry perspective, corporate governance involves the manner in which the affairs of individual banks are governed by the boards of directors and senior management, and focus on how banking institutions: (i) Set corporate objectives; (ii) Operate the bank's business on a daily basis; (iii) Meet the obligation of accountability to their shareholders and take into account the interests of other recognized stakeholders; (iv) Align corporate operations and behaviour with the expectation that banks will operate a safe and sound approach, and in compliance with applicable laws and regulations, and (v) Protect the interests of depositors.

The Bank of Tanzania has a legitimate interest in ensuring that banking institutions operate in a safe and sound manner which can only be attained if boards of directors discharge their oversight roles. The Bank of Tanzania has decided to enact these Guidelines, which spell out minimum standards required by directors of banking institutions. The objectives of these Guidelines are to (i) Promote and maintain public confidence in banking institutions; (ii) Establish standards for corporate governance processes and structures; and (iii) Guide directors for the proper discharge of their fiduciary responsibilities (Melyoki 2005).

13.3.2 Corporate Governance Practice in Publicly Listed Companies

The CMSA Guidelines state that the directors of every listed company are expected to undertake or commit themselves to adopt effective corporate governance practices as part of their continuing listing obligations. The company is supposed to underline the importance of disclosure of the extent of an organization's compliance with the Guidelines in the corporate governance annual reports and also of any areas of non-compliance or alternative practices. Further guidelines state that every listed company or issuer of securities (including debt instruments) shall disclose the compliance status. That is, whether the company is complying with the guidelines on corporate governance. Where the company is not entirely compliant the directors have to provide a detailed report on the measures the company will take to adhere to full compliance. This clarification has to be included in its annual reports and directors' report (or in the case of a debt securities issuer, the information memorandum). Furthermore, the guidelines encourage companies in the private sector other than listed corporations and issuers to practice good corporate governance.

In Tanzania and Uganda, corporate governance norms are '*principles based*' while in rest of the East African jurisdictions, corporate governance standards are a combination of '*rules based'* and *principles based'* provisions. Most of the firms in Tanzania have predominantly block ownership pattern where a few shareholders own a listed company with each one of them having a relatively large block of shares. The traditional single-tier oversight structure is applicable in the country, and the CMSA prohibits insider trading and abusive self-dealing.

Also, it is mandatory for each of the publicly listed companies to have three committees (i.e. a nomination committee, remuneration committee and audit committee), while the presence of independent directors in a majority of specialist committees is a general requirement. The audit committee consists of at least three independent non-executive directors who have knowledge about the company business and finance and audits. The existence of independent directors on the audit committee ensures the independence of the board.

As far as the composition of nomination committee is concerned, typically independent directors must account for the majority, and the chairman of the committee should also be an independent director. Nevertheless, as far as the composition of the remuneration committee is concerned the independent directors account for a majority and the chairman of the committee is also an independent director. The compensation committee consists mainly of non-executive directors.

In Tanzania, the board should comprise a balance of executive and non-executive directors. It is preferred to have a majority of non-executive directors out of which a sufficient number should be independent of management. The term of the independent directors is similar to the term of office of the directors. Each term is for a maximum period of three years. An independent director may serve consecutive terms if re-elected upon expiration of his term of office, but not exceed two terms. The CMSA have a general provision and criteria for defining independent directors whereas shareholding restrictions on and remunerations of independent directors are not specifically addressed. In addition to that, the CMSA clearly stipulates that the independent members of the Board may hold separate meetings to form an unbiased judgment on specific issues. Nevertheless, the Independent Directors have access to external legal, accounting or other specialist advice, the company's records including financial records, management, and staff. While this is optional in Tanzania, in Kenya the Act clearly stipulates that at least once in a year the committee (independent directors) shall meet with the external auditors without executive board members presence.

Regarding the mechanism of self-evaluation of the board and board members, there is no specific provision relating to the evaluation of independent members of the board in Tanzania. Whereas regarding education for the board members, it is clear that the board members will get the training necessary for orientation in the area of the Company's business, their duties and responsibilities as company board members. Nevertheless, the whole of the corporate governance norms form part of the securities commissions order whereas the Securities Commissions promote the formation and active role of Investor Associations in the place (IOSCO 2007).

Services of independent directors in Tanzania can be terminated by the shareholders at a General Meeting or by the Court upon raised complaint. The company board should receive the resignation report of independent directors immediately, and the listed company should intimate the exchange within a specified period after the departure. The board should inform the securities commissions on this resignation.

Abayo et.al (1993) examined the quality of corporate governance disclosure in the annual report for 51 Tanzanian companies. The findings revealed that the quality of corporate governance in Tanzania during this period was microscopic. On average only 52.6% of the mandatory disclosure requirements were met, with the actual levels varying across the sample from a low of 31% to a high of only 72% (Abayo et.al. 1993:152). The study used only four factors (a) the extent of compliance with mandatory disclosure requirements; (b) the extent of voluntary information disclosure; (c) the type of audit report obtained; and (d) the timeliness of the CARS. The findings suggest that even those companies that disclosed the most still failed to meet a considerable number of the mandatory disclosure requirements. But then, there were neither adequate qualified accountants and auditors nor stock exchange market in place. Other reasons specified were the lack of appropriate enforceability, high corporate governance compliance cost as compared to the low cost of noncompliance. The lack of enforceability observed during Abayo et.al's (1993) research was because by then the DSE was not established. Thus there were no CMSA in place.

In 2010 Fulgence examined the determinants of corporate governance in Tanzania. The findings revealed that corporate governance disclosure index in Tanzania (for listed companies in DSE) ranged from the lowest level of 36% to the highest level of 85.15%, giving an average of 66.58% (Fulgence 2013:47). Building on previous studies, Matinde (2013) examined the quality of corporate governance for 17 listed companies in Tanzania. His findings showed that the then overall corporate governance index (OCGI) scores of all Tanzania listed companies ranged from 51% to 74%, with an average of 62%[176] (Matinde 2013:6). In these studies, results showed that cross-listed businesses in Tanzania performed better than domestic-listed businesses in almost all categories, unlike in China where Cheung et.al.(2008) found that cross-listed companies in China only play better than domestic-listed businesses in two categories, namely the role of stakeholders and disclosure and transparency. Should cross-listed companies be excluded from the analysis, the score would significantly fall, despite this fact, the steady growth is evidenced.

13.3.3 Corporate Governance Practice in Public Sector Entities

In most African countries including Tanzania corruption was at a relatively low level during the colonial rule. However, after independence and the acceptance of single party systems, which concentrated power on small cliques, corruption began to rear its ugly head. Overtime, the environment of integrity was gradually eroded to the point that corruption in many countries in Africa is now at a level that is unimaginable (Sitta 2005). To foster effective corporate governance, create a better environment for investment and private sector development; the government of Tanzania has taken several steps. First, in 1991, the government set out the Prevention of Corruption Bureau (PCB) – now PCCB. In 1995, the government established the Public Leadership Code of Ethics to curb impropriety at higher levels of public service and later the Commission for Human Rights and Good Governance was established in 2001. Also, the government established the Permanent Commission of Inquiry in 1996 to check abuse of power by public officials and its agencies (Sitta 2005; Melyoki 2005).

When President Benjamin William Mkapa came into power in 1995, he declared war on corruption so as to enhance the environment of integrity. As a first step, he established the Presidential Commission against Corruption in January 1996. The Commission came out with one of the most respected and commended analyses of crime by any African state (commonly known as the Warioba Report) ten months later. The report identified areas and environments where corruption occurs and also revealed regulations and procedures that facilitate the crime (Sitta 2005; Melyoki 2005 & Fulgence 2014). The Warioba Report had the further benefit of opening up public discussion on corporate governance. Accordingly, the government has taken several steps to address the problem. Such measures included: - (i) Appointment of a good governance Minister, who is responsible for, among other things which are valuable in monitoring, overall strategy & implementation of anti-corruption measures. (ii) Establishment of the Commission for Ethics to deal with administering and enquiring into senior public appointee's declaration of assets and provide the recommendations to the president.

The official launch of the Public Sector Reform Programme of Tanzania by His Excellency, the President of the URT in June 2000 aimed to transform the public service into a result-oriented public service. Among other things, it aimed at creating a civil service of the high calibre and integrity that is responsive to and supportive of national efforts. Furthermore, it aimed at delivering service, to ensure effective corporate governance and to facilitate poverty reduction.

When President Dr. John Magufuli came to power in 2015, he declared a dangerous war on economic crime, corruption, organized crime and drug abuse[177]. The primary intention is to enhance transparency, the rule of law and eradication of economic crime in the region. He also aimed at turning the country and the region into an industrialized area, thereby increasing the income per capital. His slogan of '***HAPA KAZI TU***' has put more emphasis on earnest working, transparency, accountability, fairness, integrity and good governance to both public and private sector servants. In ensuring effective public fund management, he has emphasized the use of the electronic fiscal device (EFD machine) in issuing receipts. Following this, all transactions are supposed to be supported with EFD machine receipt for their authentication. The emphasis of using EFD has boosted public revenue collection. In January 2016, the Tanzania Revenue Authority (TRA) announced that a total of TZS 1.79 trillion in tax revenues were collected exclusively in that month. These results are almost in line with TRA's goal of raising at least TZS 2 trillion per month and are up by 135% from the TZS 0.76 trillion collected in January 2015[178].

13.3.4 Corporate Governance Practice in Higher Learning Institutions

In previous years, training on corporate governance was not provided in most of the higher learning institutions, due to lack of knowledge on the subject matter. In most cases, corporate governance was part of auditing theory and practice modules thus given little attention. Neither was it given the required attention in accounting professional boards like the National Board of Accountants and Auditors (NBAA) in Tanzania. Nowadays, some of the higher learning institutions in Tanzania include corporate governance training in their courses. For instance, both Corporate Governance (CG) and Corporate Social Responsibility (CSR) are amongst the taught and examined modules in most of the bachelor degree programs offered at Tanzania Institute of Accountancy (TIA). Training on Corporate Governance is in other higher learning institutions such as Mzumbe University (MU), University of Dodoma (UDOM), Ardhi University (ARU), University of Dar Es Salaam (UDSM), Institute of Accountancy Arusha (IAA) and many others. At some of the institutions, corporate governance is taught at the PG level, whilst at some others it is taught at the UG level as part of another module. These steps will ensure that graduates who will be used to steer and raise awareness of corporate governance in the market place, will be equipped with the relevant knowledge that they need.

Despite the fact that there is corporate governance training in most of the institutions of higher learning in Tanzania, there is still a problem of expertise. To date, there are very few people in the region who are fully trained on corporate governance. Even those who have had some training lack in-depth knowledge of the subject. It is no wonder that researchers on the subject matter in most of these establishments are few and far between not only in Tanzania but also in institutions of higher learning in other developed countries.[179]

13.3.5 Status Quo – Tanzania Public Sector

Within the public sector, the country took key steps and measures to promote good governance plus several reforms with effect from 2000 – such as (i) The establishment of the National Framework on Good Governance. It has spelt out critical interventions expected from the key Governance System players. (ii) Creation of the Ministry of State for Good Governance, which reports to the President's Office. (iii) The formation of the Commission for Human Rights and Good Governance. (iv) The Existence of Public Financial Management Reform Programme (PFMRP) in 2008/09 to ensure transparent accountability and institutional management as well as operational arrangements for fiscal discipline, strategic prioritization of expenses and improved performance during budget execution. (v) The review of the legal framework aimed at restoring confidence in the Judiciary and law enforcement agencies. This was accomplished through the Legal Sector Reform Program (LSRP).

There is more public focus on activities and reports from numerous oversight and watchdog institutions, including: (i) The Prevention of Corruption Bureau (PCCB); (ii) Controller and Auditor General (CAG) – National Audit Office. Robust CAG Office – oversight functions in public finance and procurement strengthened. The focus has been on capacity building. Better and favourable public opinion has increased media coverage and scrutiny of audited financial reports by the CAG. CAG's reports include recommendations to enhance effective corporate governance. Nevertheless, the Internal Auditor General's (IAG) office was created in 2011 and a circular was issued to public sector organizations in 2013 to implement Institutional Risk Management Framework. This is the renewed focus of the role of Internal Audit.

Also, in 2015, the cyber crimes law was enacted which requires enforcement at the individual and corporation levels.

Public-sector management has shown a positive trend since 2011 (Ibrahim Index of Africa Governance, 2015). Decades of public-sector reforms have resulted in Tanzania's scoring relatively better than most African countries on Public Finance Management (PFM). Procurement regulation is of international standards while transparency, oversight of the budget and revenue collection has improved. The preparations of the consolidated financial statements of public sector entities are according to the International Public Sector Accounting Standards (IPSAS) Accrual Basis. Tanzania is also a member of the Open Government Partnership (OGP), which aims to promote transparency and accountability as well as ensuring that institutions adhere to public reporting mechanisms. In his official inauguration of the 11th Parliament of the United Republic of Tanzania in Dodoma on 20th November 2015, President Dr John Magufuli promised to overhaul the government, reduce bureaucracy, introduce discipline in expenses and repeal procurement laws that created loopholes that unscrupulous civil servants had exploited for personal enrichment.

13.4 Regulators of Corporate Governance in Tanzania

Recently, Tanzania has enacted various Acts, rules, regulations, guidelines and codes for compliance with the principles of corporate governance, the enforceability of which is by the respective regulators. These regulators include but not limited to Bank of Tanzania, Dar Es Salaam Stock Exchange or Capital Market Security Authority, Business Registration and Licensing Authority, Institute of Directors Tanzania, National Board of Accountants and Auditors, Social Security Regulatory Authority, Tanzania Insurance Regulatory Authority, National Environment Management Council, etc. These regulators play a fundamental role in ensuring compliance with corporate governance requirements in their respective areas as described below.

13.4.1 The Bank of Tanzania (BoT)

Banking and Financial Institutions Act No.12 of 1991, Bank of Tanzania Act No. 5 of 2006 and Banking and Financial Institutions Regulations 2013, 2014, 2015 and many others form a cornerstone towards the supervision of banking and other financial institutions. These regulations harmonize the operations of all financial institutions in Tanzania, regulate credit services and provide for other related matters. The BoT plays a significant role in regulating banks, bureau de change and other financial institutions.

13.4.2 The Dar Es Salaam Stock Exchange/Capital Market Security Authority (DSE/CMSA)

Dar Es Salaam Stock Exchange is responsible for regulating corporate governance practices for listed companies in Tanzania. The Capital Markets and Securities Act (CMSA) No.5 of 1995, was established for the purpose of promoting the development of capital markets and securities in Tanzania and the formation of the DSE. The Act applies to listed companies in DSE. As a regulator, DSE is responsible for the enforcement of compliance with capital market regulations. The DSE and Nairobi Stock Exchange (NSE) have incorporated corporate governance guidelines into their listing requirements, following an Act incorporated in the Johannesburg Stock Exchange (JSE) after King III Report. The Report forms part of the listing rules, hence all companies listed on the JSE have to apply the rules, or explain why, if they fail to comply. The CMSA have incorporated the mandatory requirement of separating the position of the Chairman and the Chief Executive Officer (CEO) into the DSE listing rules, for all listed companies.

13.4.3 The Business Registration and Licensing Authority (BRELA)

The Business Registrations and Licensing Agency (BRELA) is an Executive Agency under the Ministry of Industry and Trade. It was established under the Government Executive Agencies Act No. 30 of 1997 and formally launched on 3rd December 1999. The agency is responsible for the administration and regulation of businesses. The authority's mission is to regulate and facilitate business operations by adopting internationally recognized best practice. All companies should comply with all rules and regulations including the Companies Act, Income Tax Act, Codes of Corporate Governance, other specific regulations required to enable the firm maintain its registration status. BRELA encourages companies to conduct seminars and educate minority shareholders about their rights. It also encourages directors to attend the training currently offered by the Institute of Directors to equip them with the knowledge and skills required to govern effectively.

13.4.4 The Institute of Directors Tanzania (IoDT)

In addition to being responsible for preparing the groundwork for the codes of corporate governance for private companies, the IoDT has established training for directors, in which they are issued with a 'Certificate of Directorship'. It is recommended that anyone requiring to be appointed into a position of directorship in a public company should possess a certificate showing that s(he) has attended the directorship course. The training is part of the initiative to equip prospective directors with the required skills to enhance their performance, so as to minimise the negative perception of directors, following major corporate scandals such as Enron, WorldCom, Parmalat, and many others (Okike,2012). This is part of the global initiatives taken in ensuring good corporate governance in various places. At the launch of the IoDT, the Vice-President of Tanzania, Dr. Mohammed Gharib Bilal, made the observation that "... There is an association of many of the problems leading to our economic crises, which resulted in poor corporate governance, with too many boards failing to meet their primary function of closely overseeing management. Most of them ended up by making decisions focusing on short term profit"[180].

13.4.5 The National Board of Accountants (NBAA)

The NBAA in Tanzania is the professional body responsible for ensuring the professional standards, skills and competences required by its members. The board provides training in corporate governance and adherence to the principles. It further enforces compliance with International Financial Reporting Standards (IFRSs), International Standards on Auditing (ISAs), as well as corporate governance principles. The board regulates both public and private sector firms and ensures that they comply with financial standards and regulations both national and international. It further regulates all professionals in the industry and in public practice and advises the heads of departments (Chief Accountants) as they assist their respective boards of directors to ensure compliance with required standards. On 31 March 2017, the NBAA issued the 'Accountants and Auditors (Practising) Byelaws', which enforces each accountant in public practice to comply with the Code of Ethics and Conduct both for NBAA and International Accounting Standards[181]. Also, NBAA issued a statement, which requires all companies to support their annual reports with a declaration issued by the Head of Finance/Accounting responsible for the preparation of financial statements[182]. The board also plays a significant role in advising the government on all matters patterning to professional accounting, auditing, and corporate governance. The board established the IoDT in May 2012, responsible for all corporate governance matters within the United Republic of Tanzania. The establishment of IoDT will enhance compliance with corporate governance codes and principles in the country.

13.4.6 The Institute of Internal Auditors (IIA)

The IIA Tanzania was formed for the purpose of creating awareness of the internal audit's contemporary role of promoting the implementation of the International Professional Practices Framework (IPPF) in Africa and advocating the globally accepted and recognized Certified Internal Audit certification (CIA) and other specialty certifications, such as the Certification in Control Self Assessment (CCSA), Certified Financial Services Auditor (CFSA), Certified Government Auditing Practitioner (CGAP), Qualification in Internal Audit (QIAL) and Certification in Risk Management Audit (CRMA). It is the sole institution in Tanzania "that provides universally applicable International Standards for the Professional Practice of Internal Auditing as well as Code of Ethics[183]". The IIA also promotes an ethical culture and professionalism in internal auditing[184], as well as improves the quality of internal auditing through the IPPF in Africa.

13.5 Current Corporate Governance Regime in Tanzania

The economic crisis in the 1980's in Tanzania was mainly caused by the paucity of effective governance of the state-owned corporations (Chachage, 2003). For instance, directors were not running the corporations for the interests of the state, but for their personal interests. As a result companies like Sungura Textile Manufacturing Industries and the City of Dar Es Salaam Transportation Company (UDA) went burst. Given the shortage of local capital to run the public corporations, the Tanzanian government embarked on the privatization of the state-owned companies (Nyaki 2013).

The method used for privatizing state-owned corporations was through 'strategic investors'. These were investors who acquired a significant number of shares effectively making them controlling shareholders. This policy provided an opportunity for some individuals to acquire significant holdings in the privatized corporations, hence the concentration of ownership of these corporations in the hands of a few individuals. Melyoki (2005) provides evidence that 52.8 per cent shares of Tanzania Breweries Limited (TBL), were held by a controlling shareholder, SABMiller Limited; 75 per cent shares of Tanzania Cigarette Company (TCC) were held by the controlling shareholder, Japan Tobacco International (JTI), and 74.14 per cent shares of Tanzania Oxygen Limited were held by another controlling shareholder, the government of the United Republic of Tanzania, thus they had incentives to exercise control.

Corporate governance in Tanzania as in other developing economies has adopted a mixture of aspects of the organizational structures found in developed markets (Tam, 1999). This situation is facilitated by legal systems, which tend to reflect the legacy of the linked colonial past between developing countries and developed western market economies (La Porta, et.al., 2000). As a former British colony, the legal system in Tanzania mirrors that of the UK (Nyaki 2013:40). The Companies Ordinance and the new Companies Act of 2002 (cap 212) contain some principles, which were applied in England in the 17th century.

Corporate governance in Tanzania is based on the Anglo-Saxon model, which reflects the liberalist approach to corporate governance. This model applies to the US, the UK, Australia, New Zealand and Canada, South Africa and other members of the Commonwealth (Bradley et.al., 1999). This model places emphasis on shareholders' interests and their power in decision-making. The fact that shareholders have invested their capital in the company gives them power over other stakeholders in managerial decision making.

The role of donor countries and the international financial institutions in the development of corporate governance practices in developing countries cannot be underestimated. The influence of the IMF and the World Bank may result in a system which does not reflect the socio-economic and local requirements for an efficient corporate governance system (Nyaki 2013:41). The preparation of the Companies Act No. 12 of 2002 relied heavily on foreign consultants and did not provide local stakeholders with the opportunity to contribute to the development of the Act.

There are internal and external corporate governance mechanisms in Tanzania. The internal mechanism is through the memorandum and articles of association (MEMARTS), which contain rules about how the company is to be governed and the duties and responsibilities of the directors and managers. The external mechanism includes legal standards and regulatory framework.

13.6 Challenges Facing Tanzania in Implementing Effective Corporate Governance

Despite several reforms and the establishment of rules and regulations to address corporate governance failures, there are still a lot of challenges to be overcome. This suggests that rules and regulations alone cannot solve all corporate governance challenges. Fulgence (2013) highlights the fact that most of the corporate collapses that occurred in countries like the UK and the US happened in spite of the existence of corporate rules and regulations. Therefore, it is not the absence of rules that matters, but compliance with the rules, and much more.

Zinkin (2010) challenged corporate governance assumptions, arguing that failures in corporate governance are performance-related, rather than compliance-related. In his book, Zinkin claimed that the destruction of these companies was because of the combination of three elements that are essential to good performance which were found to be wanting. Fulgence (2013:86) identified these factors as: (i) A lack of proper understanding by the board of its business and strategy; (ii) A total lack of appreciation of both the strategic risk and more importantly, the systematic risk created by the products' markets into which the financial institution entered because others were doing it too; (iii) A total failure of the board to ensure that the rewards and remuneration system for top executives and CEOs reflected the long-term needs of the business rather than encouraging irresponsible behaviour that placed the company's future at risk and as it turned out, the entire global financial system in jeopardy (Zinkin, 2010).

Studies undertaken by Fulgence (2013, 2014) amongst others, have highlighted areas of governance in Africa requiring attention, including strengthening parliaments, protecting the autonomy of the judiciary, improving the performance of the public sector and supporting the development of professional media. It also includes encouraging private investment and decentralizing the delivery of services. Various reports and commentaries have drawn attention to some of the main challenges to the implementation of sound corporate governance in Tanzania, including as relating to the larger economic goal of encouraging higher levels of private-sector investment. These are thought to be:

i. The overall lack of development of Tanzania's corporate sector as a result of the country's history of socialism demonstrated under path dependence theory. This problem extends to a general lack of awareness about the importance of corporate governance for economic development.
ii. The overall weak enforcement environment, which creates conditions where the concentration of share ownership has become the norm among corporations as a way to maintain shareholder control in the absence of an active market and enforcement mechanisms. The lack of the dynamic market and implementation mechanisms creates an adverse legal environment for the protection of minority shareholder interests and in turn, further, discourages the formation of the private-sector investment. Apart from banking and financial institutions, enforcement of any existing guidelines has been a challenge. Other sectors do not have oversight as close as in the banking industry.

iii. Lack of enforcement capacity among Tanzania's capital market regulators, affecting levels of accountability by corporations, directors and corporate officers and enforcement of disclosure regimes among corporations and by entities in regulated sectors.

iv. The acute shortage of qualified accountants in Tanzania, and the poor quality of professional education within Tanzania to support development and qualification of professional accountants.

v. Challenges with economic environment particularly the pervasive levels of corruption and perceived weak and ineffective judiciary, which undermines the rule of law, and thus poses a significant barrier to doing business in Tanzania (World Bank, 2016).

vi. Political influence in the constitution and operations of public sector boards which may affect their effectiveness. Also, ownership power in the constitution of private sector boards may do the same.

vii. The regulatory agenda, which is not slowing down since the increase in regulatory bodies in Tanzania. Currently, every industry is trying to establish its regulatory framework. They mostly maintain their own corporate governance guidelines. This creates grounds for conflicting requirements. For instance, listed banks are required to ensure compliance with both CMSA and BoT's guidelines. Whilst no overlaps have been identified as such, implementation is perceived as being too onerous. Also, the IoDT has established the groundwork for developing codes of corporate governance for private companies. However, this regulatory framework development is quite uncoordinated between the different regulators and may in the future result in conflicting guidelines – i.e. Zero coordination and possibility for "contagion." Global trends and developments dictate random focus improvements and responsibility across individual sectors.

viii. The unique circumstances in Tanzania makes the environment more challenging. Most of the companies are either family owned or owner-managed businesses that perceive corporate governance requirements as costly and unnecessary, since agency problems do not exist. These companies are characterized by complex control structures, secretive shareholder agreements and severe potential conflicts of interests.

ix. The board consists of nominee directors most of which are selected to board memberships on grounds other than their professional competence. Such boards further consist of very few independent directors. As a result, board independence is distorted. This is the situation which is prevalent amongst most state-owned enterprises, government companies, parastatal organizations and the executive agencies.

x. The lack of, or minimal, communication with representatives of shareholders, which prevents openness and transparency.

xi. The poor public perception of most companies, characterized by a minuscule elite business community. The recruitment into these companies is not always meritocratic and there is an obvious concern for wider social or environmental issues.

xii. The fact that board composition is made up of family members and insiders. There is no clear division between Board and Management. The board has a narrow focus on financial matters only, and there are no or ineffective sub-committees. There are informal working procedures. The board can sometimes be oblivious of important information, or could receive poor management information. Also, the oversight role for risk and control can be blurred, in addition to a lack of financial and risk 'literacy.'

The UNCTAD Study on Good Governance in Investment Promotion and Facilitation (2002), identified major obstacles to good governance in investment promotion as: (i) limited infrastructure and institutional capacity. (ii) capital resources/financial constraints. (iii) skills and education; (iv) transportation; (v) legal system and crime; (vi) bureaucracy and corruption, and (vii) institutionalised negative mind-set.

The current President, Dr John Magufuli is already taking radical steps to address these obstacles. For instance, he has stabilized the education sector by emphasizing participatory education and increasing funds for education in the national budget. The transport infrastructure is being improved through the construction of rail and road networks. The Tanzania Air Corporation is being strengthened through the provision of more aircrafts and the expansion of the airport in Dar Es Salaam, using internal funds. He is also taking steps to eradicate corruption and drug abuse.

The establishment of the Institute of Directors of Tanzania, which is tasked with the responsibility to steer corporate governance compliance is another step towards addressing these challenges. The IoDT is addressing corporate governance not only amongst listed companies regulated and guided by CSMA Guidelines on corporate compliance matters, but also non-listed companies.

13.7 Conclusion and recommendations

13.7.1 Concluding Remarks

In many African countries particularly in Tanzania, the corporate governance environment is currently being developed. To date, the findings show that while substantial progress has been made across many areas, particularly in the on-going effort surrounding private sector development, there is still significant work to be done, as overall economies are small and underdeveloped. These include but not limited to strengthening Tanzania's regulatory infrastructure and developing a sound corporate governance framework for both private and public sector enterprise. The efforts of the current president in embracing hard work, transparency, the rule of law and human rights adherence have improved the investment environment. However, further improvement in promoting effective corporate governance, working environment conducive to investment and economic growth is essential particularly in the private sector.

The private sector has an important role to play in advancing the development of Tanzania's corporate governance environment. The IoDT has been encouraged to develop a code of corporate governance for the private sector, and such a project has commenced. Tanzania has the benefit of being able to draw on experiences and learn from neighbouring countries, and other African economies as well as significant international development agency support as it continues its journey towards developing a strong economic environment.

Good governance and enhanced integrity, which are pillars for better economic and investment climate are yet to be attained in many countries. In the Corruption Index compiled by Transparency International many countries are perceived to be corrupt as the political will to counter corruption is viewed as lacking[185]. Although there is the notion that a number of countries are addressing the issue of corruption, the impact of any real change is yet to be felt. However, it appears that the new government in Tanzania is taking giant strides ahead of many others.

In the absence of high-level will within the Government, corporate governance cannot be effective (Fulgence 2012, 2014; Melyoki 2005; Nyaki 2013 & Sitta 2005). Thus it is agreed that for corporate governance to be effective, it has to be championed from the top.

In Tanzania, like other jurisdictions, there are guidelines for effective corporate governance; however, these are not yet legalized. As a result, there are still issues of effective monitoring and enforcement. The corporate governance framework is still in its infancy and needs appropriate support to make it effective. The voluntary and self-regulatory nature of the corporate governance model should be revised and harmonized across the region. Given the nature of the global business environment, corporate governance should no longer be optional.

13.7.2 Recommendations

The following recommendations are put forward, which hopefully will enhance the quality of corporate governance in Tanzania.

i. Promotion of Effective Corporate Governance at the Top Management

Senior Management mainly politicians and professionals have to institute and promote good governance and create an environment for integrity as a foundation for creating a favourable investment climate. When the government operates in an environment of integrity, it creates trust and the willingness of the business community to trust and cooperate. Such cooperation is of utmost importance in maintaining integrity (Fulgence 2014). Without strong, clear leadership and support from top government officials, its various departments and law enforcement agents, it is tough for any real advance in the fight against the evils of corruption to succeed (Sitta 2005; Fulgence 2012). Therefore corporate governance has to be at the top of the management agenda as a pro-business policy and should be embraced by all stakeholders in the society (Sitta 2005; Fulgence 2012; 2014). Finally, we need more disclosure and transparency, in both listed and non-listed companies.

ii. Promote Corporate Governance Sensitization

Promoting effective corporate governance in public awareness and adherence and emphasizing as well as sensitizing programs are needed. The information regarding corporate governance adherence and the consequences of non-compliance should be spelt out clearly in the client charters and disseminated freely (Sitta 2005; Melyoki 2005). Success in creating an environment of integrity for investment promotion and economic growth depends on the commitment and political will to implement necessary reforms as recommended (Fulgence 2012).

iii. Prevention of Resources Exploitation

Entrepreneurs in Tanzania should not exploit local human and material resources fraudulently (Sitta 2005; Melyoki 2005; Fulgence 2013; 2014). This has the potential to limit strong economic growth and development. There should be strong leaders who would protect the misallocation, misappropriation, and misuse of national recourses. For instance, the new President ordered a special audit of mining companies to establish whether they had paid their fair share of required taxes. Mining contributes 3.5% of the GDP in Tanzania, and it is the fourth-largest gold producer in Africa. Leaders like this who can dare to protect the exploitation of the country's wealth are earnestly needed in most developing countries, Tanzania in particular.

iv. Corporate Governance Research Gap

A few authors (Melyoki, 2005; Matinde, 2013; Nyaki, 2013; Fulgence, 2016 and ACGN, 2016, amongst others) have examined certain aspects of corporate governance in Tanzania. However, more research is required.

Corporate governance is now taught and examined at some of the higher education institutions in Tanzania. However, whilst training in corporate governance is a step ahead, the practical application of the principles leaves much to be desired. Besides institutions are struggling with the lack of expertise on the subject.

v. Unorganized Corporate Governance Regulators and Regulations

The multiplicity of regulations in the public and private sectors could in future give rise to overlap in, or conflicting, corporate governance requirements[186]. Efforts should be made to streamline these requirements for the benefit of various stakeholders.

REFERENCES

Abayo A.G, Adams C.A, and Roberts C.B (1993). 'Measuring the Quality of Corporate Disclosure in Less Developed Countries: The Case of Tanzania' *International Accounting, Auditing, and Taxation* 2(2):145-158.

African Corporate Governance Network (ACGN) (2016). *State of Corporate Governance in Africa: An Overview of 13 Countries*. African Corporate Governance Network, February 2016.

African Peer Review Mechanism (APRM) Panel of Eminent Persons; Country Review Report No.17: The United Republic of Tanzania; January 2013; online available at http://aprm-au.org/admin/pdfFiles/CRR_No17-Tanzania_EN.pdf as

accessed on 31st March 2017

Bradley, M., Schipani, C.A. and Sundaram, A.K. (1999) 'The Purposes and Accountability of the Corporation in Contemporary Society: Corporate Governance at Crossroads', *Law and Contemporary Problems,* 62(3): 9-86

Business Times; Tanzania advised to create corporate governance code; August 2014.

http://www.businesstimes.co.tz/index.php?option=com_content&view=article&id=3732:tanzania.advised-to-create-corporate-governance.code&catid=1:latest-news&itemid=57 as accessed on 30th March 2017

Chachage, S.L.C. (2003) 'Going Beyond Rhetoric: Why is Tanzania Still Poor 40 years after Independence?' *A Keynote Address to the Annual Convocation Symposium of the University of Dar Es Salaam.*

Cheung, YL., Ping J., Piman LY., and Tong L. (2008). Does Corporate Governance Matter in China." China Economic Review Journal 19: 460-479

Fulgence SE. (2013). Determinants of Corporate Governance Internet Reporting, LAP LAMBERT- Academic Publishing, Deutschland Germany.

Fulgence SE. (2014). 'Corporate Governance in Tanzania": In Idowu S.O and Caliyurt (Eds); *Corporate Governance; an International Perspective*. Berlin Heidelberg Springer; 157 – 185.

IMF (2016) United Republic Of Tanzania Staff Report for 2016. Article IV of Consultation and the Fourth Review under Policy Support Instrument—Debt Sustainability Analysis, 28th June 2016; online available at https://www.imf.org/external/pubs/ft/dsa/pdf/2016/dsacr16253.pdf as accessed on 19th April 2017

IOSCO 2007. Corporate Governance Practices in Emerging Markets, Report from the Emerging Markets Committee of the International Organization of Securities Commissions

La Porta, R., Florencio, L., Shleifer. A. and Vishny, R. (1999). "The quality of government", *Journal of Law, Economics, and Organization*, 15(1); 222-279.

Linder S. (2014) Tanzania: Overview of Corruption and Ant-Corruption. U4 Expert Answer. U4 Anti-Corruption Centre. March 2014; online available at http://www.u4.no/publications/overview-of-corruption-in-tanzania/ as accessed on 19th April 2017

Matinde E. (2013). The quality of corporate governance practices of Tanzania listed firms; (unpublished); A Dissertation Submitted in Partial/Fulfilment of the Requirements for Award of The Degree of Master of Science in Accounting and Finance (MSc. A&F) at Mzumbe University

Melyoki LL (2005). Determinants of Effective Corporate Governance in Tanzania, Ph.D. Thesis University of Twente, Enschede. Netherlands ISBN 90-365-2160-2

Merson, R. (2010), "*Rules Are Not Enough: The Art of Governance in the Real World*," Profiles Books Ltd, London.

Mndolwa, E.B., Binamungu, C.S. and Kashonda, M.A. (2002). 'The Companies Act No.12 of 2002 and the People's Expectattions: The Mismatch and Missed Opportunities, Accounting, Liquidation and Governance Aspects of the Law', *The Accountant, 23.*

Mo Ibrahim Foundation, 2015 Ibrahim Index of Africa Governance. Country Insight: Tanznaia.2015, online available at http://static.moibrahimfoundation.org/u/2015/10/02201457/49_Tanzania.pdf as accessed on 20th March 2017

Norad Report (2011): Joint Evaluation of Support to Anti-Corruption Efforts Tanzania Country Report, Study Report June 2011.

Nyaki, J.V. (2013) 'A Critical Analysis of Tanzanian Corporate Governance Regulation and its Impact on Foreign Investment', *Research Paper in Partial Fulfilment of the Requirments for the Master of Law Degree, Faculty of Law, University of the Western Cape.*

OECD (2014), Risk Management and Corporate Governance, Corporate Governance, OECD Publishing; Online available at http://dx.doi.org/10.1787/9789264208636-en as accessed on 18th April 2017.

Okike E. (2012). "The Importance of Education and Training in Corporate Governance and Accountability: A Nigerian case Study", *Paper presented at the 2nd African Accounting and Finance Conference,5th-7th September, Abuja, Nigeria.*

Ruparelia R., and Njuguna A. (2016) The Evolution of Corporate Governance and Consequent Domestication in Kenya. International Journal of Business and Social Science, 7(5):153-163

Sitta SJ (2005). Integrity environment and investment promotion "The case of Tanzania" (unpublished) *Paper Presented to the Conference "Alliance for Integrity – Government & Business Roles in Enhancing African Standards of Living"*; Co-organised by the OECD, NEPAD, UN Global Compact and Transparency International, in Addis Ababa on 7-8 March 2005

Sullivan JD. (2009). The Moral Campus of Companies: Business Ethics and Corporate Governance as Anti-Corruption Tools. Washington, DC: I.F.C.

Tam, O.K (1999) *The Development of Corporate Governance in China.* Edward Elgar Publishing Ltd.

United Nations Conference on Trade and Investment (UNCTAD) (2002) *Good Governance in Investment Promotion.* United Nations.

URT. 1995. Law of Ethics for Public Servants No.13 of 1995. Dar es Salaam: Government Printer

URT. 1996. Presidential Commission of Inquiry: Report on the state of corruption in the country

URT. 1999. Development Vision, 2025; President's Office, Planning, and Privatisation Commission. Dar Es Salaam: Government Printer.

URT. 2000. Privatization in Tanzania, annual review 2000/2001 and action plan 2001/2002. Dar es Salaam: PSRC.

URT. 2004. Economic survey for the year 2003. President's Office, Planning and Privatisation Commission, Dar Es Salaam

Van Lonnerenberg C. South Africa: South Africa's Developing Financial Markets Regulation Analysed. February 2013. Monaq online available at

http://www.mondaq.com/southafrica/x/222178/Financial+Services/Article+body+current+lenght+19565+chars+long+including+spaces as accessed on 18th April 2017

Warioba, JS. (2003). Remarks at the opening of the University of Dar Es Salaam Annual Convocation Symposium, Dar Es Salaam.

World Bank. (2000). *Corporate Governance: A Framework for Implementation. A report.* World Bank. (2002). *Tanzania at the turn of the century: Background papers and statistics: A World Bank Country Study*, Washington.

World Bank (2016) *Doing Business 2016: Measuring Business Regulations.* Economy Profile : Tanzania 2015.

Zinkin, J. (2010), "*Challenges in Implementing Corporate Governance: Whose Business Is It Anyway?*" John Wiley & Sons, Singapore (Asia) Pte Ltd.

Endnotes

---

[168] President's Office, Planning and Privatisation Commission (1999) Development Vision, 2025; the transition towards a multiparty political

system has also been peaceful. The stable environment provides a base for rapid economic growth.

[169] https://www.state.gov/e/eb/rls/othr/ics/2016/af/254253.htm as accessed on 19th April 2017

[170] http://www.africaneconomicoutlook.org/en/country-notes/tanzania as accessed on 18th April 2017

[171] http://www.tradingeconomics.com/tanzania/inflation-cpi as accessed on 19th April 2017

[172] https://asokoinsight.com/news/driving-africas-growth-through-fdis-inflows a

[173] http://www.tanzaniainvest.com/fdi and http://www.tic.co.tz/menu/317 as accessed on 19th April 2017

[174] http://www.bot.go.tz/BankingSupervision/BankingSupervision.asp as accessed on 19th April 2017

[175] http://www.mondaq.co.uk/article.asp?article_id=410692&signup=true as accessed on 19th April 2017

[176] The average score in Matinde's research has shown negative growth as compared to Fulgence's findings. A critical analysis indicates that this was due to the inclusion of 6 companies in Matinde's research which were newly listed businesses in DSE. The companies had very low score which had significant impact on the overall rating. Should these companies be excluded from the list, the average score would change to 68%. In this case, in can be concluded that, the corporate governance compliance in Tanzania, particularly for listed companies, is improving though steadily.

[177] http://parliament.go.tz/polis/uploads/bills/acts/1452062454-ActNo-5-2015-Book-1-10.pdf and http://www.thecitizen.co.tz/News/New-anti-drugs-boss-receives-MPs--backing/1840340-3808838-u741kk/index.html as accessed on 19th April 2017

Read more at http://www.tanzaniainvest.com/economy/tanzania-revenue-authority-tra-raises-tax-collections-by-134-percent-in-january-2016
online available at http://www.tanzaniainvest.com/economy/tanzania-

revenue-authority-tra-raises-tax-collections-by-134-percent-in-january-2016 as accessed on 20th April 2017

[179] This author had a personal experience of this shortfall in corporate governance expertise when trying to find a Ph.D. supervisor at several colleges in the US, the UK, and Australia. Most of those contacted declared that they lacked expertise in this area.

[180] Sylvester D 'Institute urged to draw up Corporate Governance Code' The Guardian 11th May 2012.

[181] Online available at http://www.nbaa.go.tz/2017/april/prbylaws17.pdf as accessed on 20th April 2017

[182]. Online available at http://www.nbaa.go.tz/declafinance.pdf as accessed on 20th April 2017

[183] The Institute Of Internal Auditors Tanzania | Progress online available at https://www.iiatanzania.org/ as accessed on 22nd April 2017

[184] http://iiatanzania.org/about/profile as accessed on 20th April 2017

[185] Transparency International. Corruption by Country: Tanzania online available at http://www.transparency.org/countryTZA as accessed on 10th April 2017

[186]. Annual reports submitted to Tanzania Revenue Authority (TRA) – for tax purposes will declare little profit; while the same annual report sent to banks or any financial institution for different purposes will state higher earnings portraying a brilliant financial position. The same applies to those presented to TIRA, TCRA, BRELA, NBAA – competing for best submitted financial statement.

Chapter 14

Corporate Governance in Tonga

Semisi Prescott

14.0 Introduction

Corporate governance has been identified as an important way forward in growing Tonga's commercial market and economy. This has been driven in part by Tonga's increasing participation in international capital markets and trade. To enable this growth, Tonga needs to attract foreign investment and maintain its credibility as a recipient of foreign aid. The Millennium Development Goals (MDG) for Tonga include economic development as a key component. The Tongan National Strategic Development Plan identifies specific targets for 2025 of increasing GDP per capita through building a more dynamic and competitive economy that is driven by the private sector. These economic targets highlight the need to introduce commercial models that are aligned with international markets. However, the implementation of these models will need to consider the social and cultural context if they are to be introduced successfully.

14.1 Background

Tonga is a constitutional monarchy located in the South Pacific with Samoa to the north, Fiji to the west and the Cook Islands to the east. The population, based on the 2011 census, is 103,252 representing a 1.2% increase compared with the 2006 census. The age distribution suggests a relatively young population with a median age of 21 years and about 55% of the population under the age of 25 years.

Although 60% of the population are at the working age (15-59) only 36% are in employment. Of those in employment 25% are in paid employment while the other 11% in subsistence work. The unemployment rate is 33% (including those in subsistence work) although the reported rate is less than 2%. The unemployment figures require interpretation given the significant proportion of subsistence agriculture. Employment income for many households is supplementary to subsistence agriculture. Of the 16,000 households living in Tonga only 30% are fully reliant on employment income. The remaining 70% primarily rely on subsistence agriculture.

The Tongan business sector is a mix of casual market traders through to established corporation. However, the market economy is relatively under developed with small locally based micro-enterprises comprising most of the aggregate market activity. The current census data does not adequately capture the number of traders and participants in this category, nor the impact that they have on the economy. It is not uncommon for households to grow much of what they need in the plantations and from fishing. The cash resources they need for expenses such as education, church responsibilities and utility supply (gas, fuel and electricity) is made possible through these micro-trading activities. Households will trade surplus produce together with hand craft items that are produced for cultural obligations and the tourist market. Remittances for extended family living abroad is another major source of income for many Tongan families (Ahlburg, 1996).

The Ministry of Commerce, Consumer Trade, Innovation and Labour maintain a list of registered companies and other entities with business licenses. Large private corporate enterprises are relatively few. Large private corporations in Tonga are often foreign owned or the result of partnerships with offshore companies and local entrepreneurs. Of the five larger hotels in Tonga, only one appears to be controlled and managed by Tongans who are resident in Tonga. The other four hotels, that are larger in size are controlled by foreign owners and directors. The true extent of foreign controlled registered corporations in Tonga is difficult to determine as many businesses that are registered as local show significant foreign ownership or have foreign resident directors with only a few who are Tongans or are resident in Tonga. There are about 200 registered businesses in the register of companies in Tonga in 2017. These companies are mainly small business start-ups in the areas of IT and communications, tourism, construction, retail, and financial and legal services. Although the business register lists about 3,000, many have been 'struck off' or are no longer operating. The age of these businesses mirrors that of many small family businesses that do not survive beyond their fifth year (Hisrich & Brush, 1986; Perricone, Earle, & Taplin, 2001; Pinfold, 2001).

Most registered corporations in Tonga are small to medium size enterprises that are typically only beginning to make the transition from being owner operated to being an enterprise that clearly separates ownership from management. Those operations that are partnerships with a foreign corporation or have independent professional directors are likely to already benefit from the advantages associated with good corporate governance. Other businesses that are at the cusp of transitioning from being a family owned and operated business to becoming a separate legal entity are faced with several challenges (Merz, Weber, & Laetz, 1994). Small business owners are faced with the challenge of maintaining sovereign control of their business while attracting the additional funding and expertise that will allow them to grow (Harvey, 2004; Vallejo, 2007; Welsch, 1993).

Many large corporations are owned or controlled by the government. These businesses are part of the public-sector network that is responsible for infrastructure, utility supply, education, health and law and order. Although these organisations are not necessarily profit motivated, they are required to maintain some degree of financial autonomy and sustainability. Tonga's relatively small commercial market and limited range of natural resources requires careful centralised coordination and management. As with many Pacific Island nations, this responsibility largely rests with the government. The governance and management of government controlled organisations is embedded in legislation and the code of conduct for public service (Code of Conduct for Public Service, 2004; Public Service Act, 2002).

Religious organisations comprise the more significant players in the philanthropic and charitable sector of the economy. Although religious organisations have traditionally played a significant role in the education sector, in recent decades it has expanded its portfolio to include commercial property and retail and social services. Management of these services has been through a hybrid governance structure that combines the traditional church hierarchy and business and financial stewardship.

The investing stakeholders that make up the current commercial market in Tonga maybe classified into four categories. The first are foreign and domestic investors who are from a western commercial background and have set up business in Tonga as an extension of existing offshore operations. These investors provide capital, commercial experience and professional knowledge to the Tongan economy. The second group are often local investors who have grown from being a micro-enterprise to becoming a registered company. These businesses are privately owned and operated and are often family based. The first and the second category share a profit driven motive. The third category are those commercial operations that are controlled by the government while the fourth category include those commercial operations controlled by religious organisations. Both the third and fourth category normally share a not-for-profit motive.

14.2 Challenges facing Tongan Commercial Businesses

Before discussing the corporate governance practices in Tonga and the benefits that can be expected to flow from good governance, it is appropriate to understand the specific challenges faced by each of the four investor groups. Category one (overseas and experienced investors) are generally familiar with western corporate governance models. The experience they have, coupled with the overseas resources and knowledge they can access, provides them with a comparative advantage. However, these investors may be regarded as foreigners and their familiarity with Tongan culture, language and the local operating environment may be obstacles. To overcome these, investors may partner with local entrepreneurs who then act as agents for the company. In some cases, these local partners may take a shareholding in the company and oversee the day-to-day management of the operations. Having a local partner may also address the investment regulations that often exist with the ownership and utility of land. These types of arrangements are not new in Tonga and there are many examples of how they have worked well in the past.

Foreign investors may also partner with the government and or with individual members of the government. Contracts with the government are potentially problematic as they stand to compromise the objectivity of the government as the chief regulating body. Instances where the partnering arrangements have been with individual members of the government provide a similar potential conflict. The objectivity of the member is potentially compromised as is the objective rule of the government.

The second category of investors is perhaps the most vulnerable in relation to corporate governance. This group often consists of local investors who have grown their business from a micro-enterprise to being a registered corporation. This group also includes minority entrepreneurs who want to set up a business as an alternative income source to employment. The relatively small size of Tonga's employment market encourages a significant proportion of the working population to follow this pathway (Hisrich & Brush, 1986). The challenges that face small to medium size enterprises and family owned business globally are similar to those for this group of investors (Cassar & Holmes, 2003; Romano, Tanewski, & Smyrnios, 2001). This group of investors is among those who may potentially benefit most from good corporate governance.

The third category of investors is the government. For small developing economies like Tonga, it is not uncommon for the government to influence the growth of the economy through its own commercial activities. Provision of essential services, infrastructure, utility supply and public goods represent a high proportion of Tonga's Gross Domestic Product. While these services are normally funded from taxation revenue, this has not been possible for Tonga. The relatively low level of taxation revenue limits the number of services that are fully funded by the government.

The deficit between government revenue and expenditure is partially managed through foreign aid and charging users. However, the accountability conditions under which foreign aid is granted create additional compliance requirements for organisations that use foreign aid to fund their operations. Since much of the foreign aid is channelled through the government, it is important that the government's processes for managing these funds have integrity and transparency. Many of the rules governing the receiving and application of foreign aid is embedded in legislation. While the policies and procedures for managing these funds are well documented, competing demands coupled with wide interpretation of the aid guidelines has attracted criticism. Good corporate governance in relation to the use of these funds is important to ensure they continue in the future.

The fourth investor category includes religious organisations. The major churches in Tonga each run a school. These semi-private schools focus on the high school level where the number of state schools does not have the capacity to cater for the number of students. Furthermore, funding for high school education is only partially funded by the government. Private schools are mainly funded through fees and contributions from the church. The Wesleyan Church in Tonga has divided the governance of the church into three separate trusts. These three trusts oversee and manage the church schools, church property and national operations. The management of the church follows a long established hierarchy that is a hybrid of the early missionary societies and its current functional and regional operations. Although the church organisational structure is different from that of corporate governance, the regular election of office holders and rotation of church ministers acts to diffuse any negative long-term distortions of power. Management of the operations and resources of the Roman Catholic Church, Seven Day Adventists and the Church of Jesus Christ and Latter Day Saints is based on the policy framework specific to these churches and observed internationally. In the last twenty to thirty years, greater compliance demands in terms of financial record keeping, taxation and human resource management have placed additional pressure on the commercial activities of the church. Although the governance structure of the churches is entrenched in tradition and religious heritage, the need to address these compliance challenges has given rise to employing skillsets outside the traditional faith based criteria.

The services provided to the public by both government and church organisations are normally not-for-profit operations. However, the need to maintain financial viability while providing services that are affordable and accessible represent potentially conflicting objectives. Reaching a balance between these objectives is an ongoing challenge that may be helped through introducing aspects of good corporate governance.

14.3 A Model of Good Corporate Governance

Good corporate governance is an organisational management hierarchy that defines and recognises the demands of the stakeholders in a market economy. Corporate governance is part of an efficient capital market model. The model necessarily assumes the market participants are driven by the motivation of wealth maximisation while minimising risk. An individual's pursuit for personal gain together with the profit objectives of a company are regulated and kept in balance by the corporate governance hierarchy.

The corporate governance model go further to define key roles and responsibilities, separation of certain duties, independent input and verification and codes of good conduct and ethics. Individual companies will often go further and reflect the principles of good governance in their governance and operating policies and procedures. Good governance has been extended to include strategic planning and control. The key functions and responsibilities identified in a corporate governance model help keep the organisation focused in its core objectives.

The authorisation boundaries and delegation of authority are roles and responsibilities defined in the organisations governance policies and deed of constitution. These often entrenched clauses in the deed of constitution prevent potential conflict of duties between officers and ensure the resources of the organisation are used or consumed in accordance with its objectives.

Good governance models need to be flexible enough to recognise the context and unique characteristics of the entity that it relates to. A set of governance policies that is too rigid is likely to be quickly abandoned by its users in favour of a more liberal interpretation.

14.4 Corporate Governance in Tonga

The corporate governance model is based on a capital market paradigm. The assumptions of profit driven individuals and corporations operating in an environment of shared information, rational choice and participant homogeneity are fundamental to the effectiveness of the model. A number of writers have challenged the validity of the traditional capital market model (Etzioni, 1988). In societies where the capital market conditions are absent or weak the benefits of good corporate governance may not be recognised. Prescott and Hooper (2009) suggest that models of western accountability do not reflect the characteristics of traditional Pacific Island societies that are based on commons. Consequently, when these models are implemented in a non-western context, the outcome is often non-compliance. Helu (1997) suggests that economic security and justice that is essential for good governance requires a degree of social cohesion that is made possible through the use of traditional customs. Although Helu was referring in this case to political governance, the notion can be extended to the successful implementation of corporate governance.

Prescott (2009) outlines that Tongan culture is based on a collective social identity where the needs of the extended family and village outweigh the needs of the individual and sometimes the immediate family. Social status is important as the collective opinions of the wider society is held in high regard. Wealth and individual profit are a means of elevating one's social status but without the social recognition personal wealth is redundant or hollow. Equally important is the collective view on sharing. Sharing is a traditional method of distribution that re-enforces relationship ties and social stature.

Effectively implementing a model of good corporate governance in Tonga will require careful recognition of its culture and customs. The social ties between families and family members, the political hierarchy and religious beliefs are inseparable components of Tongan society and life in Tonga. They are also inseparable components of the way they interact and conduct commerce both as consumers and as traders. Recognition of Tongan culture and customs may include expanding the agency rules to reflect the collective identity that Tongan people share. Agents of a company should not only act in the best interest of the firm but also in the interest of the individuals and communities they serve. The concept of profit and wealth creation and accumulation may be modified to focus on financial sustainability and continuity. This will be particularly relevant for organisations in the not-for-profit sector essential services.

14.4.1 Efforts to promote Good Corporate Governance

The need for good corporate governance has been recognised at multiple levels including government departments and services, private enterprises with offshore influence and control, and faith based organisations. The government's interactions with international organisations both public and private have made it aware of the international call for good governance. This is reflected in the structure and scope of its current acts of parliament. Government policy on codes of good conduct and anti-corruption laws provide clear guidance on good governance behaviour. The Tongan National Strategic Plan outlines several economic goals that rely on good governance frameworks. Specifically the plan identified as one of its key goals the need to "create a better governance environment" (Ministry of Finance and National Planning, 2015:8). While codes of good practice are present and potentially enforceable in public sector organisations, the same cannot be said of the private sector. Commercial operators are fragmented because those who would be responsible for implementing a model of good governance may not be aware of its benefits or have the resources to put them in place. Support to encourage the adoption of corporate governance in Tonga outside the public sector is lacking. Professional organisations in New Zealand such as the New Zealand Institute of Directors and the regional chambers of commerce provide training opportunities in corporate governance. These professional organisations are only starting to emerge in Tonga and specific training in corporate governance is not currently offered.

Church organisations continue to rely on their traditional management hierarchies to operate although there is a growing recognition that more commercial skillsets are required at the management level. Churches to date have however been slow to make significant investment in corporate governance and business skills.

The many small to medium enterprises in Tonga that are not supported by a foreign investor represent the group identified in the Millennium Development Goals (MDG) as the key to sustainable economic development. However, it is this same group who are the least informed about corporate governance and the benefits that may arise from it. A strategy that had been used by the Ministry of Social Development to encourage provider capacity in New Zealand was to offer funding for governance training. The scheme was aimed at encouraging private providers of social services to grow their capacity in both service delivery and compliance. A similar exercise may increase awareness regarding corporate governance for private businesses in Tonga. This in turn will increase business sustainability and promote growth.

14.4.2 Promoting the Adoption of Corporate Governance

Efforts to promote the adoption of good corporate governance in Tonga have largely been driven from the government. This is evidenced in the governance and operational policies that pertain to public sector services and legislation. The Code of Conduct for the Public Service (Code of Conduct for Public Service, 2004) is a detailed schedule including the purpose for the code, the underlying principles and the specific work practices and relationships that are expected from public servants and contractors working for the government. Although the framework for corporate governance has been developed, compliance and enforcing these practices is work in progress.

Faith based organisations continue to follow their traditional governance and operational models although there is growing recognition for business skills in response to increasing demands for transparency and accountability. Although the faith based version of good governance may require modification, its long history is testament that it has resilience and strengths that may benefit the design and implementation of corporate governance models in Tonga.

Enterprises that have significant overseas ownership and control are generally accustomed with western models of corporate governance. These good practices and knowledge are often not shared with the business community. The establishment of a professional body such as an institute for corporate governance may help consolidate and disseminate these good practices. Business mentors and a more prominent role for the chamber of commerce will also assist in this effort as it has in other more developed economies. These strategies may not benefit the more experienced businesses in the same way as the smaller locally owned firms but they will promote greater commercial capacity and economic growth.

Tonga's small market size coupled with many small traders suggests that a market structure based on perfect competition may not work as well as one based on cooperative partnerships and managed competition. Efforts to introduce corporate governance in Tonga may need to incorporate this social dynamic into its design and implementation.

14.4.3 Institutions that promote Corporate Governance

There are currently no specific private institutions leading the charge on corporate governance education in Tonga. The tertiary sector potentially has a role to play in this area although it requires more than simply including corporate governance in its existing business programmes. The establishment of an institute of business within organisations such as the University of the South Pacific will help advance the mindset shift required to willingly adopt western models of corporate governance. The Institute of Education is a division of the University of the South Pacific created to advance Millennium Development Goals for education in the South Pacific including Tonga. Specific courses in entrepreneurship and small business management will provide the educational foundation for sound business practices in the future. Greater business related education, business incubators and mentors are likely to assist in the MDG for economic prosperity.

Professional organisations such as the chamber of commerce, institute of chartered accountants and other emerging professional bodies will be well positioned to encourage corporate governance practices. These professional bodies in more developed economies currently provide this type of service and education to its members and stakeholders. However, the funding for these professional bodies is based on member contributions, which may be problematic for Tonga's business community given its current make-up and size. An alternative means of achieving the same functionality will be to establish a professional institution that is paid for through a business levy. All businesses will therefore be entitled to the services provided by the professional body. The Pacific Business Trust in New Zealand is an example of such a body, set up to promote pacific business formations and growth.

14.5 External Support for developing corporate governance capacity

The Millennium Development Goals (MDG) for Tonga has provided a clear framework and target for developing Tonga's economy and social structure. These goals were developed in partnership with the government of the country in question and officials from the United Nations Development Programme. The United Nations has provided good support for the development of the goals and the reported progress against the targets for each goal.

The Tonga National Strategic Development Plan (2001-2004) captures the specific initiatives within each of the MDG. Corporate governance is not a specific goal in itself, but is a necessary means for achieving the goal to "promote sustained private sector led economic growth" (Ministry of Finance and National Planning, 2015: 8). Introducing a western model of corporate governance needs to be contextualised and linked to the outcome of sustained private sector led economic development. Market participants may find it easier to accept and work towards an outcome of greater economic prosperity than to embrace (with possibly blind faith) an operational model that is unfamiliar but is likely to generate the desired outcome.

Foreign aid to assist in building corporate governance capability may be interpreted as a restrictive and irrelevant form of assistance. This type of targeted funding may be well intentioned but it is often overshadowed by the day-to-day needs of cash flow, capital investment and marketing which may be regarded as having greater priority. Furthermore, successful implementation of good corporate governance is part education and part practice. The education component is understandably the easier of the two parts. Putting theory to practice is typically an iterative process that may require further training and the help of a business mentor. Reaching a status of good corporate governance practices is a learning process that is ongoing. Staff turnover and changing social and economic circumstances require continuous improvement and review. Although governance practice may initially require external assistance, businesses will be required to take on the responsibility of maintaining it in the long term.

Business mentors have been successfully used to help SME businesses in New Zealand and Australia. A similar programme in Tonga may assist in providing Tonga's SME sector with experience and knowledge that they may not otherwise access through professional consultants. If the pool of potential business mentors is insufficient in Tonga, foreign assistance may take the form of bringing them in from overseas. It is possible that retired business entrepreneurs may appreciate the lifestyle in the Pacific Islands while providing services they have mastered during their working career.

14.6 Global Developments in Corporate Governance and their impact on Tonga

Global development in corporate governance is an inevitable reality for countries and businesses that participate in the international market. Alignment of accounting standards, taxation and transfer of wealth between countries are examples of the shift towards a common set of rules for international market participants. Although these regulations are tailored to the national agenda of each country, smaller nations including Tonga will have limited influence on their scope and impact. Regulations and best business practices may be mandated through legislation. The alternative is to encourage adoption through market forces. Corporate governance as an operating model for business is not legislated in Tonga for adoption by private businesses. Flexibility for companies to create their own constitution and operating procedures support innovation and entrepreneurship. Tonga currently follows the process of encouraging corporate governance practices but does not enforce it through legislation. Enforcing adoption of corporate governance practices may speed up the process of aligning Tonga with the global markets. However, without fully assessing the potentially negative consequences to Tongan society, culture and business sovereignty, compulsory adoption is risky.

The efforts to continue promoting corporate governance practice (or at least a modified version of it) should continue as part the Millennium Development Goals. Furthermore, this will help in preparing the Tongan economy for greater participation in the international market. The strategy for promoting corporate governance adoption should be a shared responsibility including the government, education providers, business sectors and faith based organisations. This is because the shift towards a capital market economy with western practices of corporate governance is a paradigm shift. The shift is not simply a case of adopting a new way of doing business but changing the core values and beliefs that characterise Tongan society. The process is likely to take time as shift in customs happen over long periods (Helu, 1997).

14.7 A way Forward

Developing nations in the commonwealth are faced with the challenges of developing their economy to allow for greater prosperity and sustainability. The long-term goal is to build economies that are sustainable and that are not reliant on foreign aid and remittances. The Millennium Development Goals have provided a template and targets for achieving this long-term goal. The third and final report on the MDG in 2015 identified that Tonga was making progress towards achieving its goals in terms of developing a strategic plan that captures the specific targets for each MDG domain. The targets and timeframe is optimistic. However, they provide a starting point that can be modified as new challenges emerge and others are addressed.

The specific focus of this chapter is to discuss the strategic adoption of corporate governance practices in Tonga as a means of developing economic prosperity that is driven by private businesses. The progress to date is mixed. Public sector enterprises have a documented governance structure and codes of practice that mimic practices overseas. These practices have been weaved into the operating and governance policies and procedures of the various ministries and crown entities. However, adoption at the practice level is still work in progress as on going training and mentoring is still required.

Faith based organisations are unlikely to adopt corporate governance in its pure capitalistic form. Instead they are likely to adopt those practices that they see at the time that promote the core objectives of their organisation. In favour of the faith based organisation is the underlying value and belief on which they are built. The core principles that characterise faith based organisations act as means of mitigating fraud and corruption. Regular meetings and election of officers spread the power base and the potential for corruption.

Overseas investors and entrepreneurs have assisted in the overall promotion of corporate governance practices although this is unlikely to have been part of their profit-making agenda. These have identified some of the challenges associated with running a business in Tonga. Their unfamiliarity with Tonga's commercial environment and language are barriers that have been overcome through partnering with local partners and agents. These local partners act as a conduit between the business and the local customer base and workers. It is inevitable that these agents and partners will gain hands on experience of corporate governance and western business practices.

Local investors and entrepreneurs are among the most vulnerable in terms of making the transition to a more western style of business practice. Part of the vulnerability extends from the lack of resources they have that would allow them to access professional expertise in corporate governance and other associated practices. These local entrepreneurs may benefit from a business advise bureau and or business mentor programme.

The chamber of commerce may take a lead role in capturing and making these best practices known through sponsored seminars, workshops and publications. These efforts will not only help to disseminate good practice but bring the business community together for their mutual benefit, greater awareness and social capital.

14.8 Summary and Conclusion

The process for promoting the adoption of corporate governance practices in Tonga has already been accepted as a strategy for achieving sustained economic development. However, the initial assessment as to how this should happen and the impact that it may have on Tongan society does not appear to have been appropriately evaluated. The contextual evaluation that was carried out as part of the development of the sustainable development goals and the millennium development goals correctly capture the cultural characteristics of Tongan society. Despite this, these same cultural characteristic and nuances do not appear to be reflected in the implementation design. The objectives of the United Nations to support the development of emerging economies in the Commonwealth has translated to imposing western capitalistic models with little regard to the nation's traditions and customs. Small countries who are reliant on foreign aid and international markets for imports are obliged to adopt foreign corporate practices or suffer the negative economic consequences. This need not be the case if the way forward is a co-design model that reflects the character of a country and the merits of global best practices.

As with many new projects, a good idea that is poorly implemented will have the same negative result as a bad idea that has been well executed. Good corporate governance in Tonga is a good idea that has the early indications of being poorly implemented. Small countries may openly accept business practices that promise economic prosperity but then harbour an invisible resistance because the new practices do not align with their core values. The solution is an implementation process that adopts the principles of change management. Host nations need to take ownership of the problem and the solution and then to co-design with their global partners a scheme that achieves the same outcome but incorporates the cultural nuances that help identify it as a sovereign nation.

REFERENCES

Ahlburg, D. A. (1996). Remittances and the income distribution in Tonga. *Population Research and Policy Review, 15*, 391-400.

Cassar, G., & Holmes, S. (2003). Capital Structure and the Financing of SMEs: Australian Evidence. *Accounting and Finance, 43*, pp 123-147.

Code of Conduct for Public Service, (2004).

Etzioni, A. (1988). *The Moral Dimension: Towards a New Economics*. New York: The Free Press.

Harvey, D. (2004). *Keeping it in the Family*. Retrieved from London:

Helu, F. (1997). *Tradition and Good Governance*. Discussion Paper. State Governance in Malenesia, (3). Canberra Australia.

Hisrich, R. D., & Brush, C. (1986). Characteristics of the Minority Entrepreneur. *Journal of Small Business Management, 24*(4), pp1-8.

Merz, R., Weber, P., & Laetz, V. (1994). Linking Small Business Management with Entrepreneurial Growth. *Journal of Small Business Management, 32*(4), pp. 48-60.

Ministry of Finance and National Planning. (2015). *Millennium Development Goals Final Report*. Retrieved from Tonga:

Perricone, P. J., Earle, J. R., & Taplin, I. M. (2001). Patterns of Success and Continuity in Family-Owned Businesses: Study of an Ethnic Community. *Family Business Review, XIV*(2), pp. 105-122.

Pinfold, J. F. (2001). The Expectations of New Business Founders: The New Zealand Case. *Journal of Small Business Management, 39*(3), pp 279-285.

Prescott, S. (2009). *Pacific Business Sustainability in New Zealand: A Study of Tongan Experiences.* (Doctor of Philosophy), Auckland University of Technology, Auckland.

Prescott, S., & Hooper, K. C. (2009). Commons and Anti-commons: Tongan Business Experiences in New Zealand. *Pacific Accounting Review, 21*(3).

Public Service Act, 34 Stat. (2002).

Romano, C. A., Tanewski, G. A., & Smyrnios, K. X. (2001). Capital Structure Decision Making: A Model For Family Business. *Journal of Business Venturing, 16*(3), pp 285-310.

Vallejo, M. C. (2007). Is the Culture of Family Firms Really Different? A Value-based Model for Its Survival through Generations. *Journal of Business Ethics*, pp. .

Welsch, J. H. M. (1993). The Impact of Family Ownership and Involvement on th eProcess of Management Succession. *Family Business Review, 6*(1), pp 31-54.

Chapter 15

Corporate Governance in Uganda

Juliet Wakaisuka - Isingoma and Stephen Korutaro Nkundabanyanga

15.0 A brief on Uganda

The Republic of Uganda is a land-locked country in East Africa. Uganda is bordered on the east by Kenya, on the north by South Sudan, on the west by the Democratic Republic of the Congo, on the southwest by Rwanda and on the south by Tanzania. The southern part of the country includes a substantial portion of Lake Victoria, shared with Kenya and Tanzania. Uganda gained independence from Britain on 9 October 1962. The period since then has been marked by intermittent conflicts, most recently a civil war against the Lord's Resistance Army. According to Nkundabanyanga et al. (2014) these conflicts have often had wide-ranging effects on corporate governance. Nevertheless, since the mid-1990s Uganda has experienced consistent annual increase in economic growth (6.9 per cent), higher than the African average of 5.4 per cent (World Bank, 2007), although it has slowed down due to the global economic slowdown and adverse climatic conditions affecting agricultural activities.

There are two major sources of corporate governance in Uganda: The Companies Act, 2012 and the Institute of Corporate Governance of Uganda (ICGU) Manual on Corporate Governance (2001) (www.icgu.org). Corporate governance in Uganda can be traced from the commencement of the Companies Act Cap.110,

Laws of Uganda, in 1961. For example board of directors is explicitly stated under Part V of the said Act. Among the provisions is included the number of directors, appointment and removal of directors and age limit, directors' duties and assignment of duties by directors. According to the Uganda - African Peer Review Mechanism (APRM) National Commission (2007) unitary boards are the dominant structure for public corporations, and public and private limited liability companies in Uganda. In order for both public and private companies to be registered, the Act stipulates a minimum of two directors.

Corporate governance has become important for the survival of companies and indeed of national economies in the increasingly global economy. Corporate governance is of particular concern in developing economies, where the mix of international investor capital and foreign aid is essential for economic stability and growth. Transition economies, like Uganda, are faced with the challenge of restructuring for greater efficiency and creating a foreign investment-friendly environment. In this regard, good corporate governance as a crucial facet for business practices in both public and private sectors becomes a necessity. In Uganda, the institute of Corporate Governance manages corporate governance in Uganda (ICGU). The ICGU is a not-for-profit membership based organization which was incorporated in December 1998 as a company limited by guarantee and not having a share capital, and was launched in October 2000. The vision and mission of ICGU is:

An enterprise sector that upholds international best practices in corporate governance by 2020 and promotes excellence in corporate governance principles and practices, respectively. The core values of ICGU include: Excellence, Integrity, Transparency, Accountability and Responsibility.

The key activities of the institute include;

1. Conducting and coordinating training for directors and senior management, enhancing board and management

performance, with better understanding of duties and responsibilities for the benefit of all stakeholders.
2. Publishing and disseminating information on corporate governance through public awareness lectures, pamphlets, discussion papers, the institute journal and other documents associated with the affairs and activities of the institute as well as other associated agencies.

The expected outcomes of the institute include:

1. Existence of a dynamic institute expected to create an enabling environment to foster trends in; increased business efficiency, financial probity and transparency for the benefit of all stakeholders;
2. Increased economic growth and greater trade and investment in East and Central African region; have more jobs and higher living standards; and enhance environmental sustainability

According to Kiryabwire (2014), approach to corporate governance in Uganda is broadly categorized into mandatory and voluntary frameworks. The former is where corporate governance standards are enshrined in legally enforceable instruments such as Laws and regulations and include penalties for non-compliance. This is what has come to be known as 'comply or else' regime. The latter includes codes and guidelines that contain best practices on particular corporate governance issues such as treatment of shareholders, conduct of board meetings, transparency and accountability among others. These are voluntary and do not have penalties for non-compliance. This has come to be known as 'comply or explain' regime.

Because there are both advantages and disadvantages of each approach, Uganda has adopted a hybrid approach encompassing both voluntary and mandatory frameworks. This is because such an approach draws on the strength of both and minimizes the negative effects of each. For example a mandatory framework for financial institutions under the Financial Institutions Act 2012, ensures that there is a minimum mandatory standard applicable to those categories of enterprises. Nevertheless, private companies, in particular small companies operate outside the mandatory framework and are not burdened by any onerous mandatory requirements.

The mandatory corporate governance frameworks in Uganda include the Companies Act 2012, The Financial Institutions Act of 2004 and the Microfinance Deposit Taking Institutions Act of 2003. These Laws provide certain basic corporate governance requirements that are mandatory such as the requirement to have a board of directors, to file annual returns, to hold an annual general meeting, to have a share register, to have audited accounts and allow shareholders to inspect the company's books of accounts. These are contained in the Companies Act 2012.

15.1 Efforts in Uganda to promote good standards of corporate governance and business practices in the public and private sectors, including the Codes of Best Practices

Corporate governance is perceived as important for companies. In Uganda, good governance is seen to pursue eight distinguishing characteristics: discipline, transparency, independence, accountability, responsibility, fairness, ethical conduct and good corporate citizenship. It is believed that these traits are critical for better performing organisations which are sustainable in the long-run. There is a general understanding that this in turn leads to Uganda's economic development as such organisations contribute to the economy and to society as a whole directly through the wages, salaries and taxes they pay, and indirectly through the money they pay their supply chain, the vendors, retail outlets, service and training firms and resellers of their products and services. In this regard, both public and private companies have tried their best to promote good standards of corporate governance and business practices as listed above. Public and private sectors have promoted codes of best practices in a number of ways: First, public companies have demonstrated their commitment to good governance through compliance with regulation by recognizing the importance of regulations, their application, interpretation, objective and rationale. Second, regulatory requirements are expressed through duties of directors, committees like audit committee, asset liability management committee, risk management committee, compensation committee and the auditors. Third are the remedial measures and administrative sanctions taken.

In general companies in Uganda adopt the corporate governance framework and pay attention to world-class standards on health and safety, climate change, and corporate governance underpinning environmental, social and governance considerations. A number of companies have complied with most of them broadly because regulators, customers, partners, shareholders and the public are only comfortable with such companies that demonstrate compliance with codes of best practices. Typically, best practices are benchmarked against international standards like the UK Corporate Governance Code and King III. Among the organisations that are doing a great work towards promoting good standards of corporate governance and business practices in the public sector include: the government of Uganda through Bank of Uganda (BoU), Electricity Regulatory Authority (ERA), Insurance Regulatory Authority (IRA), Ministry of Local governments (Districts and sub-county levels), and Ministry of Public Service, to mention but a few. Private companies equally work hand in hand with the government. Government of Uganda liberalised the economy by inviting the private sector to lead economic growth albeit recognising legal and institutional frameworks to allow joint investments in the economy. This would allow full exploitation of opportunities in different markets igniting these companies to improve their capacities and therefore compete effectively.

In the same thinking, Wakaisuka – Isingoma, et al. (2016) reveal that corporate governance is critical especially in financial institutions since information disclosure, financial transparency, ownership structure and board profile stimulate firm performance. It is imperative to note that in Uganda, the majority of firms are small and do not have listed shares, yet also the large state-owned, foreign owned and private family-owned companies whose shares are listed are not widely traded locally. This could explain why many people still doubt the significance of corporate governance for the economy. Broadly, the following have been crucial in promoting and demonstrating good standards of corporate governance and business practice:

15.1.1 The Companies Act 2012

Although enacted in 2012, the Act commenced on 1st July 2013 and replaced the Companies Act

Cap 110. The New Companies Act makes a bold introduction of mandatory corporate governance in Uganda. Section 14(1) of the Act provides that a public company shall at the time of registration of its articles, adopt and incorporate into its articles the provisions of the code of corporate governance contained in Table F of the Act. The manner in which section 14(1) was drafted has created some uncertainty. Some commentators have argued that the intention of the clause was to make it mandatory for all public companies to adopt Table F. However, the following wording of section 14(1) "…a public company shall at the time of registration of its articles…" appears to exempt public companies that were registered prior to the enactment of the Act. This would effectively make Table F mandatory only for new public companies. Kiryabwire (2014) suggests that this would be absurd considering that the majority of companies in Uganda are private which according to section 14 (2) may adopt the code on a voluntary basis. There are a few public companies and to make Table F only mandatory for new ones seeking to incorporate under the Act limits the application of Table F. Nevertheless for both public and private companies that adopt the code of corporate governance in Table F, it is mandatory to file a statement of compliance with the Registrar of Companies and the Capital Markets Authority by virtue of section 14(4) although there is no guidance as to the nature and content of the statement of compliance. It can be expected that the Capital Markets Authority will provide the required guidance as to the format and content of the statement.

Table F makes provision for the following on corporate governance.

a) *The Board of Directors.* The code spells out the responsibilities of the board and these include providing strategic direction, ensuring effective controls and compliance with the Laws and regulations among others.
b) *Board composition.* The code provides that the board shall be comprised of a balance of executive and non-executive directors. The code also requires the nomination committee to consist entirely of non-executive independent directors whose responsibility is to select directors in a transparent manner. Our experience suggests that this may not be practical since most boards in Uganda are dominated by representatives of shareholders and selection of board members is done taking into account various factors such as shareholder representation, skills and competencies, the desire to form strategic relationships, inter alia.
c) *The relationship between the chairperson of the board and the chief executive officer.* The code provides for a separation of roles. Nevertheless, where the two positions are combined, it provides for an independent director to be appointed as deputy chairperson.

d) *Directors*. The code categorises directors into executive, non-executive and independent. Executive directors are those involved in the day-to-day management and employed by the company or its subsidiaries. A non-executive director is one who is not an executive director. Yet an independent director is a non-executive director who does not represent a majority shareholder, has not been employed by the company in past 3 years, and is not an immediate family member of a person who is or was in the past 3 years employed in executive capacity, is not a professional advisor, is not a significant supplier or customer, has no significant contractual relationship with the group, is free from any business or other relationship which could materially interfere with his or her ability to act independently.

e) *The remuneration of director*. The code recognizes the importance of remuneration as incentive to retain quality directors. It provides for a remuneration committee comprising of independent directors and for the annual report to disclose members of the committee as well as particulars of directors' remuneration, share options and other benefits.

f) *Board meetings*. The board shall meet at least once in three months, regularly review processes and procedures as well as ensure effectiveness of internal controls.

g) *Board committees*. Board committees assist the board in the performance of its duties. Nevertheless the overall responsibility remains with the entire board. The code provides that a company should at least have an audit committee and remuneration committee and for committees to be chaired by independent non-executive directors.

h) *Board and director evaluation*. The code provides that the board should at least once a year, review its effectiveness through self-evaluation. It is considered good practice for a board to review its performance and there are various methods that are used including external evaluation and peer evaluation. Although both are commonly used, according to Kiryabwire (2014), it appears there are inherent challenges with external evaluation since the person conducting the evaluation does not attend board meetings and is not able to track director engagement and contribution in and outside the board room as well as observe board dynamics in different settings.

i) *Dealings in securities*. This is a measure to prevent insider trading and prohibits directors and officers from trading in securities of the company in the period between the end of an accounting period and the date on which accounts are published.

j) *The office of the company secretary*. The code recognizes the important role played by the company secretary in corporate governance and requires the board to empower him or her to effectively perform the duties expected of a company secretary.

k) *Risk Management*. The board has the responsibility for risk management and shall decide on risk tolerance levels and on an on-going basis, shall identify, measure and proactively manage risks. The code provides that the board shall decide on risks relating to physical and operational, human resources, technology, business continuity, credit, market, compliance and disaster recovery plans.

l) *Organisational integrity and code of ethics*. The code requires organisations to demonstrate commitment to ethical standards at the highest level and ensure adequacy of procedures to implement, monitor and enforce the standards.

m) *Accounting, Auditing and financial reporting.* The code requires financial statements to be presented in accordance with international financial reporting standards unless otherwise allowed by the Institute of Certified Public Accountants of Uganda.

n) *Relations with shareowners.* The code encourages constructive dialogue and engagement of institutional investors. It is not clear why the code focuses on relations with institutional investors only. We argue that although they play a critical finance and governance role, the board should ensure that there is constructive engagement with all shareowners including minority shareowners.

15.1.2. The Financial Institutions Act (FIA), 2004

The Financial Institutions Act, 2004 provides the regulatory framework for Banks and micro finance deposit taking institutions in Uganda. The Act defines a Bank as any company licensed to carry out financial institution business as its principal business, as specified in the Second Schedule to the Act and includes all branches and offices of that company in Uganda. Part VII contains provisions on the corporate governance requirements of commercial banks. These include appointment and removal of directors, management of conflicts of interest, conduct of board meetings, Board committees, appointment and duties of internal auditor, appointment and duties of the external auditor. In terms of conflict of interest, Section 54 of the Act requires directors and officers of financial institutions to disclose matters in which they have an interest and not to take part in discussion or take decisions on any matter in which they have interest. In terms of the audit committees, Section 59 of the Act requires the board of directors of every financial institution to constitute an audit committee consisting of non-executive members of the board. The duties of the audit committee includes reviewing internal audit reports, internal controls, procedures and management information systems, ensuring that the institution's audit function is adequate, coordinating internal and external audits, reviewing investments and transactions that may affect an institution, and monitoring insider transactions. In terms of the asset and liability management committee, Section 60 of the Act requires such a committee to primarily be concerned with risk and provision of guidelines on issues such as limits on loan to deposit ratio, loan to capital ratio, exposure to single or related customers, reliance on a particular deposit liability category, dependence on interbank and other volatile funding instruments, maximum and minimum maturities for newly acquired assets and liabilities, among others. For control over management, Section 77 of the Act gives Bank of Uganda (the central bank) powers to remove from office a chairperson, director or the chief executive of a financial institution if satisfied that it is in the public interest or for preventing the affairs of the financial institution from being conducted in a manner detrimental to the interests of the depositors or for securing the proper management of the financial institution, it is necessary to do so.

15.1.3 The Microfinance Deposit Taking Institutions Act, 2013 (MDI Act)

The MDI Act regulates microfinance deposit-taking institutions. Section 2 of the Act defines microfinance business as the acceptance of deposits, employing such deposits wholly or partly by lending or extending credit, including the provision of short term loans to small or microenterprises and low income households, usually characterized by the use of collateral substitutes, such as group guarantees or compulsory savings. Part IV of the Act contains provisions on ownership and corporate governance including the requirements for appointment and removal of directors, duties and responsibilities of the board, appointment, qualification, duties and responsibilities of the finance manager, internal and external auditors, reporting and record keeping obligations. Some of the key corporate governance requirements for

MDIs include: Limitations on institutional ownership and control, fit and proper requirements for directors, directors' duties, auditor rotation and reporting obligations on external auditors.

Regarding limitations on institutional ownership and control, Section 21(1) of the MDI Act restricts ownership of shares in a MDI to 30%. In addition, anybody who owns 10% or more of the shares in an MDI must be fit and proper according to Section 21 (4) and (5). The criteria for fit and proper as provided for in the second schedule to the Act, and it includes probity, competence, soundness of judgement, criminal and track record including any convictions for fraud, causing financial loss, engaging in fraudulent activities and defaulting on a loan of an institution in which one is a director. The FIA has similar criteria for directors and substantial shareholders contained in the third schedule to the FIA 2004. As regards directors' duties, the MDI Act not only codifies some of the traditional common Law duties of directors, such as the duty to act honestly and in good faith and the duty to act in the best interests and for the benefit of the institution, but also broadens the scope of directors' duties to include a duty to act independently, free from undue influence of any person and a duty to access necessary information to enable the director to control and discharge his or her responsibilities. In addition, to the collective responsibility of directors, the Act requires the board of directors as an organization and each director individually to immediately report in writing to the central bank if they have reason to believe that the institution may not be able to conduct its business properly as a going concern; the institution appears to be or may in the near future be unable to meet all or any of its obligations under the Act; and, the institution has suspended or is about to suspend any payment of any kind. In terms of Auditor rotation, Section 34 of the MDI Act imposes a three-year limit on the period that an external auditor can continuously serve. Under the Section 67 of the FIA 2004, this is a four-year limit for commercial Banks. When it comes to reporting obligations on external auditors, although traditionally external auditors report to the person who appoints them and with whom they have contractual relationship, the MDI Act 2013 broadens the duties of external auditors and imposes reporting obligations on them on matters such as irregularities or illegal acts and insolvency. Section 69 of the FIA 2004 contains similar provisions relating to auditors of commercial banks.

15.1.4 Voluntary Corporate Governance Frameworks in Uganda

Voluntary corporate governance frameworks include the Institute of Corporate Governance of Uganda (ICGU) Corporate Governance Guidelines of 2008, and the Capital Markets Authority (CMA) Corporate Governance Guidelines of 2003. The Corporate Governance Code in Table F of the Companies Act 2012 is voluntary for private companies.

The ICGU developed a manual in 2001, updated in 2008, that contains recommended guidelines for public and private entities in Uganda. The guidelines attempt to achieve a balance between the local environment and international standards. They cover issues such as the board of directors, internal systems and controls, the audit committee, compliance with the Laws and regulations, communication, good faith and confidentiality, code of ethics, shareholders and their duties to the employees and the community among others. The guidelines go further to define the duties of employees of an entity subject to the guidelines as well as those of external stakeholders who include the government, regulatory bodies, suppliers of a company, lenders of a company, its customers, investors and the company at large. Extending the reach of the guidelines to the above listed array of external stakeholders promises exceptional performance for corporations and other entities in Uganda that adopt them. Although the guidelines are intended to apply to most Ugandan companies and provide a basic framework for the establishment and development of a culture of corporate governance in Uganda, their adoption is voluntary. It is projected that the guidelines will be a living document that should be updated as regularly as possible in tandem with international developments as well as changes in the local context.

The CMA corporate governance guidelines were developed in 2003 in response to the growing importance of governance issues both in emerging and developing economies and in recognition of the role of good governance in corporate performance, capital mobilization and maximization of shareholder value as well as protection of investors' rights. They provide a minimum standard of good corporate governance practices by public companies and issuers of corporate debt in Uganda. The guidelines provide for: best practices relating to the board of directors; best practices relating to the position of the chairperson and chief executive; best practices relating to the rights of shareholders; best practices relating to the conduct of general meetings and best practices relating to accountability and the role of audit committees.

Other noteworthy efforts are contained in:

- Public finance management Act 2015
- African Development Plan (Chapter seven – Corporate Governance)
- Africa Peer Review Mechanism (APRM) Country Reports
- Bank of Uganda
- Corporate governance manual with recommended guidelines
- Trainings conducted by the ICGU for both public and private companies
- Private Sector Foundation
- Government through Directorate of integrity and ethics
- Inspector General of Government
- Non-government organisations like anti-Corruption coalition, Civil society and others
- The Financial Institutions (Corporate Governance) Regulations, 2005- Statutory Instruments Supplement No. 18

Largely, promoting good standards of corporate governance and business practices in the public and private sectors of the Commonwealth acts as a means to achieve global standards of efficiency, commercial probity and effective economic as well as social development. This eventually facilitates the development of appropriate institutions that advance, teach and disseminate such standards in order to strengthen, maintain and extend good practice in corporate governance.

15.2 How Uganda is attaining global standards of efficiency, commercial probity and effective economic and social development

The country is attaining the global standards of efficiency, commercial probity and effective economic and social development broadly through adherence to the legal framework, especially the company's Act regarding registration of companies, incorporation of companies and matters incidental to incorporation. In addition, Uganda Africa Peer Review Mechanism process (APRM) was formed and is managed by an independent National Governing Council (NGC), which assumed office in 2014, with the National Planning Authority (NPA) as the secretariat. Basically, the main objective is to provide leadership, guidance and direction to the Uganda APRM process, as well as awareness and sensitization of the Ugandan public. It is therefore important that all stakeholders participate in order to achieve this goal. Much as the APRM was just initiated in 2014 in Uganda, it is equally important to note that it started way back in 2002 when African leaders set forth in the African Union's Declaration on Democracy, Political, Economic and Corporate Governance. This was based on the principle of voluntary accession and invited African countries to assess themselves on the basis of their performance in four thematic areas, namely: democracy and political governance, corporate governance, economic management and governance, and socio-economic development.

Broadly, the companies in Uganda are now profiting from a level field on which to conduct their business competitively. This is basically because once there is an accepted set of standards by which corporate bodies are refereed it becomes easier for individual corporations to formulate their own standards of behaviour. Corporate governance has recognized and upheld legitimate interests of all stakeholders to a company like shareholders, board of directors, management, employees, bankers, creditors, to mention but a few. Corporate governance has been documented at political levels through standards of transparency, accountability, probity and integrity in the political and administrative spheres. Corporate governance has also been acknowledged in the sustainability and long-term wealth creation of enterprise. In addition, there has been documentation of sound societal norms that emphasise communal responsibility through, for example, revitalization of kingdoms. Therefore, Uganda, like any other economy in transition, now increasingly recognizes that corporate governance is an essential tool for prosperity and economic growth as well as improved operational and strategic competitiveness of Commonwealth enterprises operating in the global market. Largely, Uganda cannot shield itself from the global movement that is shaping standard principles governing corporations.

15.3 Institutions advancing, teaching and disseminating corporate governance standards in Uganda

There is a considerable amount of work that has been advanced, taught and disseminated by a number of institutions in promoting good standards of corporate governance and business practices in Uganda. These include:

- Universities like Makerere University Business School and Uganda Martyrs Nkozi
  These universities have a good working relationship with

ICGU. Additionally, good practices of governance are embedded in their teaching curriculum but also, students and lecturers, subscribe to ICGU as members: therefore do attend conferences, use different platforms to promote codes of best practices in corporate governance to society as a whole.

- Association of Chartered Certified Accountants (ACCA)
  ACCA disseminates corporate governance through, governance, risk and ethics that aims at making the recipient apply relevant knowledge, skills and exercise professional judgment in carrying out the role of the accountant relating to governance, internal control, compliance and management risk within an organization in the context of an overall ethical framework. ACCA therefore portrays the relevance of corporate governance in an organization setting, given its core objectives, among which are; first, it defines governance and also explains its function in the effective management and control of organisations, as well as, the resources for which they are accountable. Second, it evaluates the professional accountant's role in internal control, review and compliance. Third, it explains the role of the accountant in identifying and assessing risk. Fourth, it explains and evaluates the role of the accountant in controlling and mitigating risk. Finally, it demonstrates the application of professional values and judgement through an ethical framework that is in the best interests of society and the professional, in compliance with the relevant professional codes, laws and regulations (Kaplan Financial Limited, 2015).
- Institute of Certified Public Accountants in Uganda
  The Institute of Certified Public Accountants in Uganda (ICPAU) follows the Accountant Act 1992 with the aim to regulate and maintain standard of accountability in the country and prescribe or regulate the conduct of accountants

in Uganda (www.icpau.co.ug). The institute is also governed by a council that is assisted by two statutory committees - the examination board, and the disciplinary and ethics committee, together with seven more including: education committee, events management committee, technical committee, members' services committee, finance and administration committee and lastly planning and development committee. Accordingly, ICPAU maintain a disciplinary and ethics mechanism that is helpful to both the employees and members to act professionally and more so, in the interest of the employer, accountant and in public interest.

- Public Sector Foundation
  The experience over the years clearly shows that successful privatizations and the development of vibrant private sector depend to a significant extent on the existence of effective systems of corporate governance.
- Institute of Corporate Governance in Uganda
  Has based its practices on three main principles including; the Commonwealth Association for Corporate Governance (CACG) principles, the OECD principles and the Code of corporate practices and conduct (The King's Committee Report of South Africa).
- Uganda National Bureau of Standards (UNBS)
  UNBS promotes formulation and promotion of the use of standards by administering them in protection of the public health and safety as well as the environment against dangerous, counterfeit and substandard products. This enhances fairness in trade and precision in industry through reliable and measurable systems thus strengthening Uganda, as an economy. This further promotes competitiveness of local industries and therefore improvement in the quality of exports through standardization, quality assurance, testing and metrology.

- Capital Market Authority
  That is committed to practise the highest level of corporate governance and also conducts its affairs in line with the principles set out in the capital market corporate governance guidelines 2003, and the international best practices in corporation governance. Equally, it is responsible for regulating and promoting a fair, transparent and efficient capital markets industry in Uganda so as to protect its investors' interests.
- Uganda Securities Exchange
  Increased probity, efficiency and effectiveness of the securities exchange, increases direct and portfolio investor confidence to commit long-term funds to the country. This requires establishing a comprehensive set of best practices to which listed companies should adhere. If it is considered to be in the best interests of the company not to follow one or more of these standards, the company should disclose this to its shareholders, along with the reasons for not doing so. Otherwise, below are the guidelines of the Capital Markets Corporate Governance, where Uganda Securities Exchange (USE), subscribes.

The Capital Markets Corporate Governance Guidelines (2003) embeds the following:

a) The authority developed the minimum standards for good corporate governance practices by public companies and issuers of corporate debt in Uganda in relation to the growing importance of governance issues in the emerging economies as well as promoting domestic and regional capital markets growth. To this regard also, is the recognition of the role of good governance in corporate performance, capital formation and maximization of shareholders value as well as protection of investors' rights.

b) In line with the guidelines, corporate governance involves the process and structure used to direct and manage business affairs of the company towards enhancing prosperity and corporate accounting with the aim of protecting and promoting shareholders' rights, realizing shareholders long-term value as well as taking into account the interests of stakeholders.
c) The guidelines are equally in line with several works that have been undertaken extensively in several jurisdictions through many task forces and committees among which are, though not limited to the United Kingdom, Malaysia, South Africa, the Commonwealth Association for Corporate Governance and the OECD principles of corporate governance.
d) The Authority also adopted both the prescriptive and non-prescriptive approaches so as to provide for flexibility and innovative dynamism to corporate governance practices by public listed companies.
e) Good corporate governance practices are nurtured and encouraged to evolve as a matter of best practices. In addition, certain aspects of operation in a body corporate must of necessity require minimum standards of good governance. To this regard therefore, the Authority expects the directors of every listed company to commit themselves to adopt good corporate governance practices as part of their continuing listing obligations.
f) In the same reasoning, the Authority has supported and has been useful in the development of a code of best practices for corporate governance in Uganda, which has been issued by the ICGU, and therefore supplementary thereto.
g) Finally, the ultimate objective of these guidelines is to strengthen corporate governance practices by listed companies, in Uganda, and promote the standards of self-regulation in order to bring the level of governance in line with international trends.

- Bank of Uganda- Specifically the Financial sector
  The financial sector is assigned the special position of trust held by financial institutions in Ugandan economy and their access to government welfare nets, therefore obliged to have strong corporate governance mechanisms. Second, with the increasing globalisation of financial markets, emergence of conglomerate structures, technological advances and innovations in financial products, require that the quality of corporate governance in financial institutions should be reasonably high. Third, weak corporate governance is a basic cause of bank failures in Uganda (about 10 banks have closed down: 1996 to date) and therefore needs to be strengthened (The Financial Institutions (Corporate Governance) Regulations, 2005). The substance of this attention certainly springs from the crisis, which became a dominant reminder of the importance of the financial system in any economy, as a healthy economy cannot exist without a well-functioning financial system.

15.4 How the transition and emerging economies within the Commonwealth have been supported to strengthen their capacities to maintain and extend good practice in corporate governance

Since the financial crisis in 2008, policies have been promoted, proposed, debated, and endorsed on virtually every aspect regarding corporate governance. There is growing evidence that corporate governance is creating steadiness between economic, social, individual and communal goals while encouraging the efficient use of resources, accountability, the use of power and stewardship as well as aligning the interests of individuals, corporations and society (Hartono, Subroto, Djumahir & Irianto, 2013). The financial crisis was therefore a good reminder that many countries are operating below the required standards of good practices in corporate governance. To this regard, in order to strengthen capacities to maintain and extend good practices in corporate governance, the following are some of the examples that have been implemented in Uganda: partnerships with corporate governance sister companies, African corporate governance institutions, Companies Act, Laws coming up, for instance, are incorporating corporate governance, corporate governance charters as well as training and write-ups in corporate governance. In implementing corporate governance strategies in the Commonwealth, the CACG designs a comprehensive package of activities, which act as a platform that ensures the process not only conforms with standards of international best practice, but also, that it is directly linked to national priorities and concerns.

15.5 The impact of global developments in corporate governance on the development of corporate governance in Uganda

Uganda has made significant strides in corporate governance. However, there are still a number of shortcomings that need to be addressed. For example, some laws are not fully aligned with international standards and political interference in the administration of justice, compounded by pervasive corruption, undermines the Ugandan investment climate. Corporate financial transparency is largely lacking with the exception of banks, insurance, and listed companies. Also, despite commendable progress made in strengthening systems for commercial and labour dispute resolution, the judiciary remains plagued with backlogs. In addition, the supervisory and regulatory bodies face critical shortages of human, technical, and financial resources, which render them ineffective. In this context, a number of private institutions have been created (i.e., the Institute of Directors, the Shareholders' Association, etc.) with a view to improve accountability and strengthen corporations and their directors and management. Notwithstanding, these institutions still have low acceptance and lack sufficient resources to integrate corporate governance adoption. Similarly, there is need to also expand the institutions' coverage to include public awareness regarding the importance of corporate governance in Uganda. This could be the reason why African countries rank highest on the sub-index and director liability sub-index (African Development Report, 2011). Nonetheless, various Acts have been introduced to strengthen capacity and maintenance; donor pressure (in partnerships) has emerged and therefore extending good practices in corporate governance, technology through information flow as well as World Bank guidelines in the different fields and departments. In addition, the International Finance Corporation (IFC) –the World Bank's private arm is funding private sector projects in Africa, Uganda inclusive, and broadly considers state of corporate governance before the funding is provided.

REFERENCES

Adotey Bing – Pappoe (2010). *Reviewing Africa's Peer Review Mechanism: A Seven Country Survey*, Partnership Africa Canada, ISBN 1-897320- 17-5.

African Development Report. 2011. Corporate Governance - Chapter seven: 133 – 148

Corporate Governance Manual (2008). Incorporating recommended guidelines for Uganda. ICGU. United Printers Ltd, Kampala – Uganda.

Hartono, U., Subroto, B., Djumahir, G., & Irianto, G. (2013). Firm characteristics, corporate governance and firm value. *International Journal of Business and Behavioural Sciences,* 3, (8), 9 -18.

http://www.icgu.org [Accessed 20/08/2016]

http:/www.icpau.org [Accessed 20/08/2016]

Kaplan Financial Limited. (2015). *Governance, Risk and Ethics-Paper 1*: Kaplan Publishing, Wokingham, Berkshire, UK.

Kiryabwire, TW. (2014). 'The Legal and Regulatory Framework for Corporate Governance in Uganda'. In Katto J., Wanyama S and Musaali, EM. (editors), *Corporate Governance in Uganda: An Introduction to Concepts and Principles*, Fountain Publishers, Kampala, 69-85.

Nkundabanyanga K. S. Ahiauzu A., Sejjaaka S., and Ntayi MJ. 2014. 'Intellectual Capital in Ugandan service firms as mediator of board governance and firm performance'. *African Journal of Economic and Management Studies,* 5 (3): 300-340.

The Capital Markets Corporate Governance Guidelines (2003): Section 102 of the Capital Markets Authority Statute, 1996; Statute No. 1 of 1996.

The Financial Institutions (Corporate Governance) Regulations. (2005). Statutory Instrument No. 18: Statutory Instrument Supplement, Uganda Printing and Publishing Company, Entebbe, Uganda.

Uganda APRM National Commission. (2007). The Uganda Country Self-assessment Report and Uganda Bureau of Statistics. 2016. The National Population and Housing Census 2014 – Main Report, Kampala, Uganda.

Wakaisuka – Isingoma J., Aduda J., Wainaina G., and Iraya CM. 2016. Corporate governance, firm characteristics, external environment and firm performance of financial institutions in Uganda - A Review of Literature. *Cogent Business and Management*, 3 (1): 1 -14.

World Bank (2007) https://openknowledge.worldbank.org/.../Uganda000Count0valuation00200102007.t...100005 [Accessed 16/08/2016]

Chapter 16

Corporate Governance in the United Kingdom

Elewechi Okike

16.0 Introduction

The United Kingdom is an island nation in North Western Europe. It is a sovereign state and a constitutional monarchy, with the Queen as the ceremonial head. The Prime Minister, Theresa May, exercises executive power on behalf of the Monarch. It is a leading financial centre for the rest of the world and boasts a position of one of the largest economies of the world.

The UK is an important member of the Commonwealth of Nations, more so, as the Queen is the Head of the Commonwealth. Appropriately, the secretariat of the Commonwealth organisation is based in London, the capital of England. The city played host to the most recent Commonwealth Heads of Government Meeting (CHOGM), which took place in London, in April 2018. It is one of the developed countries in the Commonwealth and consists of four countries, England, Scotland, Wales and Northern Island. It currently has a population of around 66 million (based on estimates from the Office for National Statistics)[187].

In the 19th and early 20th centuries, when the British Empire spread across different continents, all the countries that were under British rule had to adopt British company law. Whilst the company laws of many of these countries have developed over the years to reflect their particular socio-political and cultural environments, many of them still have some semblance of British company law, especially through their association with the Commonwealth.

16.1 Corporate Governance and Company Regulation

Whilst company law does not provide a comprehensive framework for corporate governance, nevertheless some elements of company law are essential in helping us to understand the key relationships in the corporate world.

UK company law draws from both statute and common law. All companies incorporated in the UK are subject to statutory regulation, irrespective of their size. However public interest companies are subject to greater legislative requirements, including additional regulation within their specific industries.

The law recognises a limited company as a separate legal entity. Their Memorandum of Association and Articles of Association dictate internal management of companies. Boards derive their powers from their Articles of Association, or constitution. The Articles of Association make provision for directors to delegate their powers to board committees and executive directors. Whilst directors have powers to act as executive directors and members of the board, they also have duties to act as responsible managers of the companies in which they are employed.

Directors are agents of their companies and consequently have certain duties and obligations to the company, as a legal entity, and not to individual shareholders, employees or any other third party outside of the company. Prior to the introduction of the provisions of Companies Act 2006, the main duties of directors were duties in common law, meaning a fiduciary duty to the company, and not to its shareholders. However, Companies Act 2006 has written these common law duties into statute law. Sections 171-177 of the Act reveal what these duties entail:

S.171 - Duty to act within powers
S. 172 - Duty to promote the success of the company
S. 173 - Duty to exercise independent judgment
S. 174 - Duty to exercise reasonable care, skill and diligence
S. 175 - Duty to avoid conflicts of interest
S. 176 - Duty not to accept benefits from third parties
S. 177 - Duty to declare interest in proposed transaction or arrangement

The common law duties apply to both executive and non-executive directors.

Ironically, whilst shareholders collectively own the company, their powers within the law are limited. However, in the context of corporate governance, shareholders can take actions, which can affect decisions taken by directors on behalf of the company.

The most significant powers of shareholders is in their power to vote at the annual general meeting in relation to issues such as the election and re-election of directors, appointment and re-appointment of external auditors, approving or reducing the proposed dividend and decisions on authorised share capital of the company.

In addition to their limited powers, shareholders also have certain rights, including the right to receive a copy of the annual report and accounts of the company and the right to attend and vote at the annual general meetings of the company.

Directors make themselves accountable to the shareholders through the annual report (including the interim financial statements). The annual report is important in relation to corporate governance because it is the channel of communication between directors and shareholders. The document contains a mixture of information required by statute, accounting standards, the Combined Code and others voluntarily provided by the directors.

Section 471 of the Companies Act 2006 details what the "annual accounts and reports" of a quoted company for a financial year should include:

(a) its annual accounts
(b) the directors' remuneration report
    (ba) the strategic report (if any)[188]
(c) the directors' report, and
(d) the auditor's report on those accounts, on the auditable part of the directors' remuneration report, on the strategic report (where this is covered by the auditor's report)[189] and on the directors' report.

Section 292 of The Act requires that the financial accounts of a company give a true and fair view of the state of the financial affairs of the company and must be prepared in accordance with international accounting standards (Sec. 395)

Section 417 of The Act requires that the directors' report include a business review describing how the business developed during the financial year, the nature of any risks or uncertainties it faces, any trends that may affect its future development, an analysis of key performance indicators and also sections in relation to employees and other social and environmental issues. The directors' report should also include details of any gains the directors have made whilst exercising their options, income received from long-term incentive plans, severance payments and other benefits.

Section 495 of The Act requires that auditors must state in their report which financial statements were audited and what standards were used in the audit process. They must state whether or not the financial statements were prepared in accordance with relevant accounting standards and with the Act, and whether or not they give a true and fair view of the company's affairs. Their report must also state whether in their opinion the information given in the strategic report (if any) and the directors' report are consistent with the annual accounts, and whether any such strategic report and the directors' report have been prepared in accordance with applicable legal requirements (Sec.496).

16.2 The Financial Reporting Council and Company Regulation

The Financial Reporting Council (FRC) was established to help restore the confidence of investors in corporate governance following the spate of corporate scandals, in the UK and elsewhere. It was established as an offshoot of the numerous committees that had been set up after Cadbury 1992 to promote transparency and integrity in business.

The FRC is responsible for setting the UK Corporate Governance and Stewardship Codes. It also regulates auditors, accountants and actuaries. Prior to the Consultation that took place in October 2011, the FRC operated under the following structure: The Accounting Standards Board (ASB), the Auditing Practices Board (APB), the Board for Actuarial Standards (BAS), the Professional Oversight Board (POB) which includes the Audit Inspection Unit (AIU), the Financial Reporting Review Panel (FRRP), and the Accountancy and Actuarial Discipline Board (AADB). However, following the consultation that took place in October 2011 and the responses received, the government introduced legislation restructuring the FRC, with effect from 2 July 2012. The FRC powers originally vested in the six operating bodies is now devolved to the FRC Board, which oversees a much-streamlined structure.

The Board is supported by three governance committees[190]: Audit Committee, Nominations Committee and Remuneration Committee; two business committees: Codes & Standards Committee and Conduct Committee, and three advisory councils: Corporate Reporting, Audit & Assurance and Actuarial. The Corporate Reporting Review (CRR) Committee, Audit Quality Review (AQR) Committee and the Case Management Committee support the Conduct Committee and have specific responsibilities as set out in the FRC's monitoring, review and disciplinary procedures. The Financial Reporting Review Panel (FRRP) and the disciplinary Tribunal Panel are maintained pursuant to the Conduct Committee Operating procedures and the FRC's Disciplinary Schemes.

What protection does the law offer to shareholders against poor corporate governance? The next section examines the system of corporate governance in the UK.

16.3 System of Corporate Governance

The U.K. adopts an Anglo-Saxon or outsider-dominated system of corporate governance in which ownership of corporate equity is dispersed amongst a vast number of outside investors/shareholders. In this type of system, large firms are controlled by managers/directors (also known as 'agents') but owned predominantly by outside shareholders (often referred to as the 'principals'). Such a system of corporate governance faces agency problems (Jensen & Meckling 1976; Shleifer & Vishny, 199); Bonazzi & Islam, 2007) because of the separation of ownership from control. This system is also prone to hostile takeovers (Weisbach, 1993; Sullivan & Wong, 2005), which act as a mechanism for disciplining company management. Although ownership is dispersed, the large range of shareholders, which are often institutional investors, exercise some degree of control over management. Investors enjoy strong protection in company law, and this system of corporate governance has potential for shareholder democracy (Engelen, 2002; Enriques & Volpin, 2007).

16.4 Development of Corporate Governance Codes

Corporate governance issues emerged in the UK as a result of highly published corporate scandals in the late 1989s such as Polly

Peck and Maxwell giving this historic issue a new thrust. The corporate problems which arose, consisted of many financial reporting irregularities, creative accounting, surprising business failures, the auditors limited role and inconsistent link between directors remuneration and company performance, to mention just this few (Agrawal & Cooper, 2017). Subsequently, the first corporate governance committee, the Cadbury Committee was formed in 1991 in an attempt to rectify concerns over corporate scandals, resulting in 1992 to the publication of the Cadbury Report.

16.4.1 Cadbury Report (1992)

The Financial Reporting Council, the Stock Exchange and the accountancy profession set up the Corporate Governance Committee in May 1991, in response to continuing concern about the standards of financial reporting and accountability, particularly in light of the BCCI and Maxwell cases.

The Committee was chaired by Sir Adrian Cadbury and had a remit to review those aspects of corporate governance relating to financial reporting and accountability. The final report, *The Financial Aspects of Corporate Governance* (known as the Cadbury Report) was published in December 1992 and contained a number of recommendations to raise standards of corporate governance.

16.4.2 Greenbury Report (1995)

In the 1990s as investor confidence was diminishing as a result of the scandals that had erupted in the corporate world, the issue of directors remunerating themselves above what was considered reasonable became a topical matter. More so, investors could not see any link between company performance and directors' remuneration.

In January 1995 the Confederation of British Industry (CBI) established the Study Group on Directors' Remuneration under the chairmanship of Sir Richard Greenbury with a remit to identify good practice in determining directors' remuneration and to prepare a code of practice for UK PLCs. The final report of the group was published on 17 July 1995 and is usually referred to as the Greenbury Report.

16.4.3 Hampel Report (1998)

The Committee on Corporate Governance (the Hampel Committee) was set in November 1995 to review the Cadbury Committee's recommendations on corporate governance and the Greenbury Report. The Hampel Committee released a preliminary report in August 1997, followed by a final report in January 1998. It suggested that the recommendations of all three committees be integrated into a single code of corporate governance.

16.4.4. Combined Code 1998

With the publication of the Hampel Report in 1998, the UK already had three reports looking at various aspects of corporate governance. It was therefore decided that the recommendations of Cadbury, Greenbury and Hampel be accepted and consolidated into a single report, from which it derived its name, *The Combined Code*. The Code consisted of two sections, one aimed at companies, the other aimed at institutional investors. Following its publication in 1998, the Code was adopted by the London Stock Exchange and included in the appendix of the UK listing rules.

The Code established standards of best practices for companies in relation to the composition of their boards, directors' remuneration, accountability and audit. It required companies to have sound system of internal controls in place to safeguard shareholders' investments, and directors were to ensure these systems were reviewed regularly to ensure their continued effectiveness.

The Code required directors to state in their annual reports whether or not they had complied with the Code, and if they had not, whether they provided an explanation to that effect; hence the 'comply or explain' approach adopted in the UK. It is an approach to corporate governance that is 'principles-based', in comparison with the 'rules-based' approach adopted in the US.

16.4.5 Turnbull Report (1999, revised 2005, 2014)

In order to maximise returns for their shareholders, board of directors cannot avoid taking certain risks. Managing such risks and having adequate systems of internal controls in place to mitigate such risks are important. Appropriately, a Working Party of the Institute of Chartered Accountants in England and Wales (ICAEW), the Turnbull Committee, chaired by Nigel Turnbull, was established to provide guidance on how companies were to implement the internal control requirements of the Combined Code, including compliance, financial, operational, and risk management. The Committee produced the Turnbull Report, which was first published in 1999 and set out best practice on internal control for UK listed companies.

The Turnbull Guidance was fully reviewed in 2004-2005, and in October 2005 the Financial Reporting Council (FRC) issued an updated version of the guidance with the title '*Internal Control: Guidance for Directors on the Combined Code*'. Although there were no significant changes to the scope and content of the guidance, it was instead amended to encourage more informative disclosure. The guidance required 'boards to confirm that necessary action has been taken to remedy any significant failings or weaknesses identified from the annual review'.

This guidance was again superseded in September 2014 by the *Guidance on Risk Management, Internal Control and Related*

Financial and Business Reporting to reflect changes made to the UK Corporate Governance Code. Its aim was to 'bring together elements of best practice for risk management; prompt boards to consider how to discharge their responsibilities in relation to the existing and emerging principal risks faced by the company; reflect sound business practice, whereby risk management and internal control are embedded in the business process by which a company pursues its objectives; and highlight related responsibilities' (FRC, 2014).

16.4.6 Myners Report (2001, 2008)

The Myners Report - *Institutional Investment in the UK: A Review*, was a report that looked at institutional investment in the UK and established a best practice approach to investment decision making for pension funds. It was a report submitted to HM Treasury in March 2001. The report suggested that institutional shareholders should be more proactive in their role and ensure that shareholders were getting the best out of their investments. Following the publication of the results of the National Association of Pension Funds (NAPF) review of compliance with the Principles recommended in the Myners Report, HM Treasury and the Department for Works and Pensions issued a response, *Updating the Myners Principles: A Response to Consultation* in 2008.

16.4.7 Higgs Report (2003)

The Higgs Committee was charged with reviewing the role and effectiveness of non-executive directors. It was chaired by Derek Higgs, and submitted its report in January 2003. The aim was to develop guidelines for making them more effective in fulfilling their role. NEDs needed to understand their role, be knowledgeable, experienced and know the company and its affairs reasonably well. They should have enough time to perform their duties and have the opportunity to contribute to board decision-making.

16.4.8 Tyson Report (2003)

A task force commissioned by the Department of Trade and Industry, led by Dean Laura D'Andrea Tyson of London Business School submitted a report, *The Tyson Report on the Recruitment and Development of Non-Executive Directors* in June 2003. The report followed the publication of the Higgs Review of the Role and Effectiveness of NEDs in January 2003. The report highlighted how a range of board members from different background and experiences can enhance board effectiveness. The report contained recommendations on strategies for recruiting and developing NEDs.

16.4.9 Smith Report (2003), revised 2008)

The Financial Reporting Council (FRC) Group on Audit Committees, chaired by Sir Robert Smith, was tasked with developing the guidance on audit committees in the Combined Code. The group's report, commonly referred to as the Smith Report, was published on 20 January 2003 and codified the role of audit committees. The purpose of the report was to provide guidance to company boards on how they are to put in place adequate arrangements for their audit committees and for individual members of the audit committees on how best to fulfil their role and responsibilities. The report was subsequently revised and is now known as the FRC Guidance on Audit Committees.

16.4.10 Combined Code (2003)

This Code was published in June 2003, following the recommendations made in the Higgs Report, and was a revised version of the 1998 Combined Code. It incorporated the recommendations of the Higgs Committee, which proposed (FRC, 2003) that:

- at least half of a board (excluding the Chair) be comprised of NEDs;
- the NEDs should meet at least once a year in isolation to discuss company performance;
- a senior independent director be nominated and made available for shareholders to express any concerns to; and
- potential non-executive directors should satisfy themselves that they possess the knowledge, experience, skills and time to carry out their duties with due diligence.

Other recommendations include the need for separation of the roles of chairman and the chief executive officer; stating the number of meetings of the board, including the attendance records of individual directors in the annual report. Chairmen and chief executives should ensure resources were in place for the training and induction of NEDs; no one NED should sit on the three major committees (audit, nomination and remuneration) of the board.

16.4.11 Combined Code (2006)

The FRC made changes to the earlier (2003) version of the Combined Code following two consultation exercises. The first consultation held between July and October 2005 was to assess the overall impact of the 2003 Code, and the second consultation, which was between January and April 2006 was on the draft amendments to the Code. The revised Code was published in June 2006 and applied to financial years beginning on or after 1 November 2006.

16.4.12 Combined Code (2008)

Following consultations in 2007, the FRC issued a revised version of the Combined Code to reflect EU requirements in relation to

Audit Committees and corporate governance statements. There were also changes in relation to the appointment of board chairman and membership of the audit committee.

Turner Review March (2009)

The financial crisis caused mayhem across the global business community. As a result the government was keen to identify the cause of the crisis and what needed to be done to avoid a repeat of such in the future. Appropriately, the Chancellor of the Exchequer commissioned the Chairman of the Financial Services Authority (FSA), to carry out this exercise and make recommendations. His recommendations focused on the need for the banking sector to be "a shock absorber in the economy, and not a shock amplifier". He described his recommendations as 'profound, and the banking system of the future will be different from that of the last decade. The world's economy will be better served as a result'.

16.4.13 Walker Review (2009)

The financial crisis caused mayhem across the global business community, especially within the financial sector. As a result the UK government commissioned Sir David Walker to examine corporate governance in UK banks and other financial industry entities and come up with some recommendations. His review covered the role and constitution of board of directors; the size and composition of the board, the role of institutional investors, how banks managed their risks, and the remuneration of directors, and qualification of Board members; the functioning of the board and the evaluation of their performance. The preliminary conclusions and recommendations were published in a consultation document on 16 July 2009, and the final report was published on 26 November 2009.

His recommendations were comprehensive, having taken into consideration the comments received during the period of consultation. In relation to the board size, composition and qualification, the report suggested that non-executive directors needed to commit more time to their roles in the banking and other financial institutions (BOFI). They required proper training to enable them carry out their responsibilities effectively. The board of BOFI must be well balanced with members having appropriate skills and experience in the sector. The chairman of the BOFI needs to devote at least two-thirds of his/her time to the work of the board, and must have the experience and skill necessary for the role. There should be a rigorous evaluation of the performance of the board and the relationship between the board and institutional shareholders should be more consolidated and formalised through the Stewardship Code. The report suggests that ‘the board of a FTSE 100-listed bank or life insurance company should establish a board risk committee separately from the audit committee’. The board risk committee will be responsible for advising the board about the various potential risks, taking into consideration all macroeconomic and microeconomics issues and the overall financial environment. The board risk committee should submit a separate report, which should be included in the annual report and accounts. Sir Walker recommended that the remuneration committee should have sufficient understanding of the company’s approach to pay and employment conditions to ensure that its approach is commensurate with the work undertaken by bank employees. In relation to “high end” employees, ‘the remuneration committee report should confirm that the committee is satisfied with the way in which performance objectives and risk adjustments are reflected in the compensation structures..’ For FTSE 100-listed banks and comparable unlisted entities, the remuneration committee report should disclose in bands, the number of “high-end” employees and executive board members whose total expected remuneration in the year under report is in a range of £1 million to £2.5 million, in a range of £2.5 million to £5 million and in £5 million bands thereafter (showing within each band the components of salary, cash bonus, deferred shares, performance-related long-term awards and pension contribution). Implementation of this recommendation, amongst others, should strengthen accountability within this sector.

16.4.14 UK Corporate Governance Code 2010

Following the financial crisis of 2008 this Code became applicable for reporting periods beginning on or after 29 June 2010 and applies to companies with Premium Listing of equity shares, whether or not registered in the UK. Like previous Codes, the main principles centre around leadership, board effectiveness, accountability, remuneration and relations with shareholders. The Code suggests that it is the responsibility of the board to determine the nature and the extent of the risks it is willing to take to achieve its strategic objectives.

16.4.15 The UK Stewardship Code (2010) (revised 2012)

A credible investment climate is crucial to the success of the UK economy, especially in the highly competitive global market. Institutional investors are key participants in UK corporate governance. In order to ensure they engage meaningfully and strategically with the company and its management, the Stewardship Code was instituted. The Code aims to improve the quality of communication between the company and the institutional investors and sets out good practice on engagement to ensure dialogue is meaningful and addresses issues that could be contentious at annual meetings, including management of risks, performance and corporate strategy. The idea is to build a critical mass of UK and overseas investors, who are keen to engage in high quality dialogue with companies and help create a stronger link between governance and the investment process. It would form the basis for complying with the Code or for providing an explanation for non-compliance. The FRC sees the Stewardship Code as complementary to the UK Corporate Governance Code, and it is addressed firstly to firms who manage assets on behalf of institutional shareholders (e.g. pension funds, insurance companies, investment trusts and other collective investment vehicles).

Institutional investors can choose whether or not to engage with their investee companies, depending on their investment approach. Disclosures made by institutions under the Code help companies to better understand the approach and expectations of their major shareholders, and also those issuing mandates to fund managers are better informed. These disclosures help in the efficient functioning of the market and improving relations between companies and their institutional shareholders.

Institutions are to apply the Code on a 'comply or explain' basis and provide a statement on their website that contains:

- a description of how the principles of the Code have been applied, and

- Disclosure of specific information listed under principles 1,5,6 and 7; or
- An explanation if these elements of the Code have not been complied with.

The FRC revised the Stewardship Code in September 2012 following consultations in April 2012 and taking into consideration all the responses received. The minor changes to the Code include:

- The need for companies to provide clarification of the stewardship responsibilities of asset managers and asset owners, and what stewardship activities are being outsourced.
- The requirement for investors to explain how they manage conflict of interest, the circumstances in which they will

participate in collective engagement, and how they make use of the services of proxy voting agencies.

- The need for asset managers to have their stewardship responsibilities independently verified to provide greater assurance to their clients.

The Code sets out areas of best practices, which the FRC believes institutions should aspire to achieve. Whilst announcing the changes, the FRC Chairman, Baroness Hogg explained[191],

> "The changes to the UK Corporate Governance Code are designed to give investors greater insight into what company boards and audit committees are doing to promote their interests, and to provide them with a better basis for engagement. The changes to the Stewardship Code are designed to give companies and savers a better understanding of how signatories to the Code are exercising their stewardship responsibilities".

Since December 2010 the Financial Conduct Authority's (FCA) Conduct of Business Rules requires all UK authorised Asset Managers to produce a statement of commitment to the UK Stewardship Code or explain why it is not appropriate for their business model.

16.4.16 UK Corporate Governance Code (2012)

The FRC updates the UK Corporate Governance (UKCG) Code every couple of years, the last revision being in 2010. Following a period of short consultation, the FRC issued the UKCG Code 2012, in which minor changes were made to both the Code and the Stewardship Code. The application of both codes still remains the 'comply or explain' basis. The minor changes to the UKCG Code include:

- Attempts to improve the quality and effectiveness of the external audit function, by requiring FTSE 350 companies to put out their external audit contract on tender every 10 years. Incumbent audit firms can also compete in the bid
- The requirement that audit committees provide information about how they carried out their responsibilities and how they assessed the effectiveness of the audit process
- The need for boards to confirm that their annual reports and accounts are fair, balanced and understandable, and that there was consistency between the narrative parts of their reports and the information contained in their financial statements, and reflect the company's performance.
- The requirement for companies to explain and report on the progress made in relation to their policies on diversity in the boardroom.
- The need for companies to fully explain to shareholders any deviation from the provision of the Code.

Companies are required to apply the Code from 1 October 2012.

16.4.17 UK Corporate Governance Code (2014)

On 17 September 2014, the FRC issued an updated version of the UKCG Code, following a period of extensive consultation and based on the feedback received. The Code significantly improves the quality of information that shareholders receive about the long-term health and financial strategy of their investee companies, and pays greater attention to risk management, and aligning remuneration with sustained value creation.

The following are the key changes to the Code:

Going concern, risk management and internal control

- Companies are required to state whether they consider the appropriateness of adopting a going concern basis of accounting and they should identify any material uncertainties that might preclude them from continuing to do so;
- The need for companies to assess the principal risks they face and how they are being managed or mitigated;
- Companies are required to state whether they have the ability to continue operation and meet their current obligations, taking into account their current position and principal risks, and to specify the period to which their statement relates, and why they feel this is appropriate.
- Companies should monitor their risk management and internal control systems, review the effectiveness of these systems at least annually and report the outcome of this review in their annual report.
- Companies can choose in what part of the annual report they would like to put their risk and viability disclosures. If placed within the Strategic Report, the directors will be covered by the 'safe harbour' provisions in Companies Act 2006[192].

Remuneration

- The Remuneration Committee must ensure that remuneration policies are designed with long-term survival of the company in mind;
- Companies are to put in place arrangements that will enable them to recover or withhold variable pay when it is appropriate to do so. They should also consider appropriate vesting and holding period for deferred remuneration.

Shareholder Engagement

- When publishing the results of their annual general meeting, companies should explain their strategy for shareholder engagement especially after majority of them have voted against a resolution.

In addition, the FRC made the point that it is the responsibility of the board to set a good example of culture and behaviour within the organisation. They should also encourage constructive and challenging dialogue within the board and encourage diversity in terms of their approach and experience.

Along with the UKCG 2014, the FRC also issued three related documents[193]:

i) *Guidance on Risk Management and Internal Control and Related Financial and Business* Reporting (the Risk Guidance) – an amalgamation of Turnbull 2005 and 2009 Going Concern notes and incorporating the requirements of UKCG 2014;

ii) *Guidance for Directors of Banks on Solvency and Liquidity Risk Management and the Going Concern Basis of Accounting*

iii) *Revised Auditing Standards (extracts) – ISAs (UK and Ireland) 260, 570 and 700* – requiring auditors to report on narrative disclosures, including risks; meaning they will need to consider the going concern basis of accounting and the longer term viability statement and the risk management governance.

16.4.18 UK Corporate Governance Code (2016)

The FRC issued an updated version of the UKCG Code in April 2016, with minimal changes. The changes, which affect auditing and ethical standards and company law, saw the implementation of the EU Audit Regulation and Directive, including an updated Guidance on Audit Committees. These were effective on or after 17 June 2016.

In relation to the changes in the 2016 Code, the Executive Director of Audit, Melanie Mclaren, said[194],

> "The updates to the Code, Guidance and Standards implement a significant change in audit regulation in the UK which will be overseen by the FRC as a competent authority with the support of the accountancy professional bodies. The Changes will support further innovation by the audit profession in the UK, and ensure that auditors act in a way that is genuinely independent and seen to be in the public interest. The UK has led the way on promoting audit transparency and competition on quality so that investors can have confidence in corporate reporting"

Whilst the changes to the Code itself were minimal, there were more substantial changes to the Guidance on Audit Committees and they cover audit committee activities and reporting. It is now a requirement for audit committee members to have competence relevant to the sector in which the company operates. The provision relating to the need to tender the external audit every 10 years has been removed, as it has been superseded by the Competition and Markets Authority (CMA) and EU requirements for mandatory tendering and audit firm rotation.

The Code requires companies to disclose in the audit committee report how the committee has assessed the effectiveness of the external auditor, the approach taken to the external auditor's appointment or reappointment and the length of tenure of the current audit firm.

There were also some key changes to auditing standards, including the requirement for 'enhanced audit reporting' for all listed companies and public interest entities (PIEs)[195]. The contents of the enhanced audit reports would be substantially more, as auditors would be required to include an expansion of the description of key audit risks and how they responded to those risks. Their report would also include a description of how their audit was considered capable of detecting irregularities and fraud. Other disclosures in the report include the tenure of the auditor, previous reappointments and renewals of appointment, and a declaration of the auditor's independence, and a confirmation that no prohibited services were provided.

16.4.19 UK Corporate Governance Code (2018)

The FRC has released a new Corporate Governance Code in the UK, which it describes as 'A Code Fit for the Future'. The new Code was released on 16 July 2018. According to the FRC, the new Code 'puts the relationship between companies, shareholders and stakeholders at the heart of long-term sustainable growth in the UK economy'. The Code applies to all premium-listed companies for the financial years beginning on or after 1 January 2019.

In this new Code, the FRC took on board the UK corporate governance framework recommended in the Government's August 2017 Response Paper, and builds on the findings from the FRC's Culture Report published in 2016. The FRC also issued a completely reworked version of its *Guidance on Board Effectiveness*, which replaces the 2011 version.

There are a few key new requirements of the 2018 UKCG Code. These include:

- An enhanced focus on corporate culture. In other words, companies are to align their strategy and values with culture
- Having a board-monitored mechanism for whistleblowing
- Engagement with stakeholders and the disclosure of section 172 of Companies Act 2006 requirements
- The need for all boards to have a mechanism for engaging with their workforce
- The requirement that board chairs do not stay for longer than 9 years from the date of their appointment
- Having a greater focus on gender, social and ethnic diversity in succession planning at board and senior management levels.
- Reporting on diversity for board and senior management
- An enhanced role for the remuneration committee to oversee company-wide remuneration policies
- The requirements for executive remuneration reporting
- Vesting and holding periods for long-term incentives should be at least 5 years.
- Removal of some concessions for companies outside of FTSE 350.

In terms of its structure, the 2018 UKCG Code is more robust and compact. Compared to the 99 principles, supporting principles and provisions of the old Codes, the 2018 Code contains 18 principles and 41 provisions. The remit of the Code includes all premium-listed companies.

Instead of the five sections containing a main principle, sub-principles and provisions (A-E), in the 2016 Code, the 2018 Code has five sections (1 – 5), with each section containing principles (A-R) followed by provisions, numbered sequentially from 1 to 41. Most of the supporting principles in the 2016 Code have been moved to the Guidance. The 2018 Code retains the 'comply or explain' principle.

Sir Win Bischoff, Chairman of the FRC, said[196],

> "The UK is globally renowned for its corporate governance framework, which is underpinned by the UK Corporate Governance Code. At this critical time as the country approaches Brexit, a revised Code will be essential to restoring trust in business, attracting investment and ensuring the long-term success of companies for members and wider society".

He stated further that

> "A Principle promoting the importance of the intrinsic value of corporate culture is a new addition to the Code. Building trust in business has to start in the organisation and forming a healthy corporate culture is integral to the credibility of a company. Engaging with and contributing to wider society must not be seen as a tick-box exercise but imperative to building confidence among stakeholders and in turn the long-term success of a company".

The 2018 UKCG Code would enable companies to report how their governance structure contributes to long-term success and achieves wider objectives. The revised *Guidance on Board Effectiveness* supports the new Code.

16. 5 Summary and Conclusion

In the light of the different corporate scandals that swept across Europe and the United States of America in the 1980s and the 1990s, the importance of effective corporate governance is not in doubt. The UK has been at the forefront of corporate governance reforms especially following Cadbury 1992. Since then, the corporate environment has been evolving, and the quality of corporate governance in the UK has been greatly enhanced and globally acknowledged. Investors are attracted to UK listed companies because of the trust and confidence the Code engenders (FRC, 2017). The Code's 'comply or explain' approach significantly, has allowed the UK to respond positively and effectively to evolving market circumstances, which hard rules often cannot (Biscoff, 2017). It is over 25 years ago that the Cadbury Report was published, and there is evidence (Grant Thorton, 2017) of improvements in the quality of corporate governance reporting in the UK, including an increase in the level of compliance with various aspects of the Code of Corporate Governance.

Whilst it is understood that corporate governance institutions cannot be perceived in isolation from the governance culture in which they are embedded, nevertheless, developments in corporate governance in the UK have had an impact on the corporate governance practices in other countries.

REFERENCES

Agrawal, A. & Cooper, T. (2017) "Corporate Governance Consequences of Accounting Scandals: Evidence from Top Management, CFO and Auditor Turnover", *Quaterly Journal of Finance* (7)1.

Bonazzi, L. & Islam, S. (2007) "Agency Theory and Corporate Governance: A Study of the Effectiveness of Board in their Monitoring of the CEO", *Journal of Modelling in Management*, 2(1): 7-23.

Cadbury Report (1992) *Report of the Committee on the Financial Aspects of Corporate Governance: the Code of*

Best Practice, Gee professional Publishing, London, December.

Combined Code (1998), The London Stock Exchange Limited, London, June.

Combined Code on Corporate Governance (2003), Financial Reporting Council, London, July.

Combined Code on Corporate Governance (2006), Financial Reporting Council, London, June.

Combined Code on Corporate Governance (2008), Financial Reporting Council, London, June.

Enriques, L & Volpin, P. (2007) "Corporate Governance Reforms in Continental Europe", *Journal of Economic Perspectives,* 21(1): 117-140.

Financial Reporting Council (FRC) (2017) "25th Anniversary of the UK Corporate Governance Code"

https://www.frc.org.uk/directors/corporate-governance-and-stewardship/uk-corporate-governance-code/25th-anniversary-of-the-uk-corporate-governance-code.. Accessed 15/08/2018

Greenbury Report (July 1995) *Directors' Remuneration (Report of a Study Group chaired by Sir Richard Greenbury),* Gee Professional Publishing, London.

Hampel Report (1998) *The Final Report,* The Committee on Corporate Governance. Gee Professional Publishing, London.

Holmstrom, B. & Kaplan, S. (2003) "The State of US Corporate Governance: What's Right and What's Wrong?" *Journal of Applied Corporate Finance, 15(3): 8-20.*

Jensen, M. & Meckling, W. (1976) "Theory of the Firm: Managerial Behavior, Agency Costs and Ownership Structure", *Journal of Financial Economics,* 3(4): 305-360.

Shleifer, A. & Vishny, R. (1997) "A Survey of Corporate Governance", *The Journal of Finance*, 52:737-783.

Smith Report (2003) *Audit Committees: A Report and Proposed Guidance*, Financial Reporting Council, London, January.

Sullivan, N. & Wong, P. (2005) "The Governance Role of Takeovers" in Keasey, K., Thompson, S. & Wright, M. (eds.) *Corporate Governance: Accountability, Enterprise and International Comparisons*, John Wiley & Sons Ltd., 155-181.

Thornton, Grant (2017) *Corporate Governance Review,* https://www.grantthornton.co.uk/globalassets/1.-member-firms/united-kingdom https://www.grantthornton.co.uk/globalassets/1.-member-firms/united-kingdom/pdf/publication/corporate-governance-review-2017.pdf

Turnbull Report (1999) *Internal Control: Guidance for Directors on the Combined Code,* Institute of Chartered Accountants in England and Wales, London, September.

Turner Review (2009) *The Turner Review: A regulatory Response to the Global Banking Crisis*, The Financial Services Authority, London, March.

Tyson Report (2003) *The Tyson Report on the Recruitment and Development of Non-Executive Directors*, report commissioned by the Department of Trade & Industry following the publication of the Higgs review of the role and effectiveness of non-executive directors in January 2003, London Business School, June.

UK Corporate Governance (2010), Financial Reporting Council, London, June.

UK Corporate Governance (2012), Financial Reporting Council, London, September.

UK Corporate Governance (2014), Financial Reporting Council, London, September.

UK Corporate Governance (2016), Financial Reporting Council, London, April.

UK Corporate Governance (2018), Financial Reporting Council, London, July.

Walker Review (2009) *A Review of Corporate Governance in UK Banks and Other Financial Industry Entities*, The Walker Review Secretariat, London, 16 July.

Weisbach, M. (1993) "Corporate Governance and Hostile Takeovers" *Journal of Accounting and Economics* 16(1-3): 199-208.

Endnotes

---

[187] https://www.ons.gov.uk/aboutus/transparencyandgovernance/freedomofinformationfoi/ukpopulation2017. Accessed 18/08/2018

[188] S.471(3)(ba) inserted (with effect in accordance with reg.1(4) of the amending S.I.) by The Companies Act (Strategic Report and Directors' Report) Regulations 2013 (S.I. 2013/1970), reg. 1(2)(3), Sch. para. 18(b)(I).

[189] S.471(3)(d) inserted (with effect in accordance with reg. 1(4) of the amending S.I) by The Companies Act 2006 (Strategic Report and Directors' Report) Regulations 2013 (S.I. 2013/1970), reg. 1(2)(3), Sch. para. 18(b)(ii)

[190] https://www.frc.org.uk/about-the-frc/structure-of-the-frc. Accessed 21/08/2018

[191] https://www.frc.org.uk/news/september-2012/frc-publishes-updates-to-uk-corporate-governance-code. Accessed 21/08/2018

[192] Section 463 of the 2006 Act introduces a new safe harbour in relation to directors' liability for the directors' report (which includes the business review), the directors' remuneration report and summary financial statements. Directors are only liable to compensate the company for any

loss it suffers as a result of any untrue or misleading statement in, or omission from, such a report if the untrue or misleading statement is made deliberately or recklessly, or the omission amounts to dishonest concealment of a material fact.
This safe harbour addresses the concern of directors over liability for negligence when making, for example, forward-looking statements in the directors' report, in particular, the business review. The directors' liability is limited to the company rather than to third parties.

[193] https://www.frc.org.uk/news/september-2012/frc-publishes-updates-to-uk-corporate-governance-code. Accessed 25/08/2017

[194] https://www.frc.org.uk/news/april-2016/revised-uk-corporate-governance-code. Accessed 25/08/2018

[195] A PIE is defined in EU law as an entity governed by the law of a Member State with securities traded on an EEA regulated market, a credit institution or insurance company.

[196] https://www.frc.org.uk/news/december-2017/a-sharper-uk-corporate-governance-code. Accessed 28/08/2018

Chapter 17

Corporate Governance in Lake Chad

Fadeke Ayoola

17.0 A brief introduction of Lake Chad and bordering countries.

Lake Chad, positioned on the periphery of the Sahara Desert, is recognised by the Ramsar Convention on wetlands (Ramsar, 2006). The Ramsar Convention's mission is the conservation and wise use of all wetlands, including lakes, rivers, swamps, marshes, and wet grasslands. It promotes the wise use of all wetlands through local and national actions and international cooperation towards achieving sustainable development, and classified as a World Natural Heritage Site by UNESCO (Ramsar, 2006). It is an ancient lake surrounded by four countries[197]: Chad is situated to the east of the lake; the population of Chad can access almost half of the body of water. Niger is positioned to the northwest of the lake; the population can access less than twenty per cent of the body of water. Nigeria is located to the west with access to one quarter of the body of water, and finally, Cameroon is situated to the south, with less than ten per cent access to the body of water (Joint Environmental Audit Report, 2014). This lake was previously one of the largest bodies of fresh water on the African continent. Natural causes such as droughts, water scarcity due to the advancement of desert and climate change, as well as deforestation, and various other human actions, have significantly contributed to the shrinking of Lake Chad.

Lake Chad's present state of shrinkage, influences the economy, the environment and the populations living in the Lake Chad basin. However, due to a lack of data, the present scarcity in fresh water can only be regarded as passing, rather than permanent (Odada, et.al, 2006). With adequate investment, Lake Chad could become a region for prosperity in trade and economic development for its local inhabitants, combating poverty and social unrest amongst its inhabitants. Investment could also ensure adequate data is available to improve decision-making on the future prospects of the region, as well as improve the monitoring of water levels and usage. Agriculture, animal husbandry and fishing are considered to be the prime activities in this region.

Only 38% of the surface of Lake Chad is currently water, the remaining 62% of the Lake is now made up of reeds and islands (Odada et.al, 2006). Without continued investment, to improve and strengthen its governance, environmental management and data monitoring, the present state of the Lake will profoundly affect the local biodiversity through the loss of habitats, such as fish and plant communities, affecting millions of birds, a number of mammals, reptiles and amphibians. 30 million people rely on the resources the lake provides; without continued investment their lives will be devastated.

The recent joint environmental audit report of the Supreme State Audit services of Cameroon, Chad, Niger, and Nigeria, over a three-year period 2011-2014, found serious corporate governance failures in the management of Lake Chad by the Lake Chad Basin Commission (Joint Environmental Audit Report, 2014). Firstly, the auditors found evidence that the Lake Chad Basin Commission executive secretariat invested in programmes that did not fall within its mandate or its role as basin organization, notably investing in poverty related issues instead of the lake. Secondly, the auditors evidenced a lack of coordination of policy and strategy between actors in water resource management. Thirdly, the auditor evidenced a lack of clarity in the allocation of roles and responsibilities of the member states in the execution of the strategic action programme. And finally, the awareness of the laws, legislation and enforcement regarding water resource protection in terms of dissemination, popularization and awareness raising among users was either inadequate or non-existent. Other factors, identified by the auditors included a lack of financial, human and material resources. Resources such as these are deemed necessary to acquire investment to address these key governance failures. Overall, serious governance failures needed to be addressed in order to reinforce and re-instate overall good corporate governance.

In the 1960's, after African countries gained their independence from western colonial powers, the Lake Chad Basin Commission was formed; two out of the four countries are members of the Commonwealth of Nations, Nigeria joined in 1960 and Cameroon in 1995.

The Lake Chad Basin Commission is a legitimate established official body, formed on May 22, 1964 by the Fort Lamy Convention, with the power to command three key objectives. Firstly, to govern the sustainable and equitable management of the Lake Chad waters and other trans-boundary water resources; secondly to uphold and safeguard the ecosystems of the catchment area; and thirdly, to advocate the integration and preservation of peace and security in the conventional basin. (Joint Environmental Audit Report, 2014)

The structure of the Lake Chad Basin Commission consists of three main functional areas, (1) the Summit Heads of State and Government; (2) the Council of Ministers and (3) the Executive Secretariat. The Summit Heads of State and Government, meet once a year in an ordinary conference, forming part of the political structure, assigned with the task of decision-making and planning strategic policies and guidelines. Within the Council of Ministers, is the technical committee, which also meets once a year in an ordinary session, within this session, each member state Government department is represented by a specialist to handle issues related to Lake Chad Basin such as water resource management and other issues communicated through technical documents. For Cameroon, the member state representing them is the Ministry of Economy, Planning and Regional Development, (MINEPAT). The MINEPAT coordinates development actions in trans-boundary basin areas, including the Lake Chad basin area. Niger, is represented by the Ministry of Hydraulics and Sanitation. This ministry conceives, plans, implements and evaluates water and environmental policies throughout the national territory in collaboration with other ministries and territorial authorities. The Federal Ministry of Water Resources represents Nigeria. This federal ministry formulates national policies on water resource management and ensures that it is implemented. Finally, Chad is represented by the Ministry of Hydraulics and Animal Husbandry, which executes government policy regarding hydraulic infrastructure, knowledge, monitoring, use of water resources, meteorology and husbandry. The technical committee was formed in the year 2000, to harmonize policies and regulations relative to water and the environment in the Member States. Finally, the Executive Secretariat that forms part of the executive body of the Lake Chad Basin Commission ensures that the principles issued in the Fort Lamy Convention are adhered to, and applied. The Executive Secretariat also maintains close contact with member states, conducting baseline studies and working in the Lake basin, promoting regional cooperation and the settlement of disputes, as well as planning, coordinating, and conducting regional projects. The Executive Secretariat also ensures that projects prepared by member states are considered, and if appropriate, a recommendation is issued with an action plan for the further development of the basin (Joint Environmental Audit Report, 2014).

17.1 Efforts to promote good standards of corporate governance and business practices within Lake Basin Management.

Firstly, it is necessary to acknowledge the important role of the Commonwealth in natural resource management, and to discuss what role it has played in influencing and promoting good standards of corporate governance and business practice for its members, (Nigeria, 1960 and Cameroon, 1995). The Commonwealth Secretariat supports its member countries to progress towards sustainable management and regulation of their natural resources, with a particular focus on oil, gas, renewable energy and mining. Lake Chad Basin has been explored for oil and gas by the Nigerian National Petroleum Corporation (NNPC, 2017), utilised for renewable energy in fish production by the Global Environment Facility (GEF) and the African Development Fund (ADB), who worked with the local fishing sector over a ten-year period, where riparian countries benefited from the fish production which improved income generation as well as the promotion and transfer of knowledge and technology. Lake Chad Basin has also been utilised for mining by various corporations over a number of years including the China National Petroleum Corp. The important role of the Commonwealth whose expertise span three decades, is to strengthen policies, legal and fiscal frameworks and national institutions providing advice and guidance to boost investment and enhance good governance in natural resource management. The Commonwealth also advises member states to become part of the Extractive Industries Transparency Initiative (EITI), which is a global coalition of governments, companies and civil society working together to develop and advance transparency, integrity and openness to promote accountability of the income generated from natural resources.

The complexity of water resource management within the Lake Chad Basin requires, in addition to the support of the Commonwealth Secretariat, other frameworks that have been developed to address the complexities of this particular natural resource. These complexities include watershed boundaries that infrequently coincide with political jurisdictions as well as the lack of integration of the wider components of the water system, in terms of the integration of natural resources within the ecosystem (Mackenzie, 1996).

There are also other unique characteristics of lake basin management that are different from other water bodies. For example, lakes integrate nature, have long retention time, and have complex response dynamics, other water bodies have these individual characteristics; however, lakes have all three-combined making their management challenging (Radik, 2009).

Four specific frameworks have been developed to aid state governments and other actors in the governance of natural resources, such as lakes; firstly, the Integrated Water Resource Management, (IWRM); secondly, Adaptive Management; thirdly, the Integrated Lake Basin Management (ILBM) and finally, Environmental Impact Assessments (EIA). The issue and concept of good governance in water management has only been realised at the beginning of the 21st Century. Research into this concept is still emerging. There is evidence in a recent research study that there is a need for detailed case studies to fully understand the complexity of water management, to inform theory building and future policy on the issue (Hubschen, 2011). The current water crisis has unravelled inherent limitations of institutions in their inability to adequately supply resource allocation, governance and conflict resolution mechanisms (Dinar, et.al. 2005). The established viewpoint of water as a free good is substituted with a viewpoint of water as an economic and social good. Water users are now classified as clients and consumers. According to the United Nations Development Program (UNDP), the Integrated Water Resource Management (IWRM) is defined as an interdependent governance system of implementation, planning, and decision-making for watersheds, rivers, lakes, wetlands, coastal zones, and oceans. Other definitions include, a practice or system which supports the governance, advancement and management of water, land and related resources to make the best of economic and social welfare (GWP, 2000). According to the World Water Forum, the Integrated Water Resource Management (IWRM) is a global model for water resource management. The South African Department of Water Affairs and Forestry (DWAF) defined it as a philosophy, a regional, national and international management strategy for all stakeholders. The need for an integrated water management system is critical since a global water scarcity crisis, is only one aspect of the problem, a water governance crisis, is but another fundamental issue, particularly in developing countries (Hubschen, 2011). According to the UNESCO good water governance is the solution, however, the responsible governing institutions often fail in their duty to allocate water supplies. The importance of improving this failure is critical, since the economic consequences are dire for humanity, wild life and the eco-system. There is evidence that 70% of all water is used by agriculture. A serious water shortage will have fundamental impact on food security, wild life and the eco-system. Practitioners, specialists and consultants around the globe have deemed IWRM, as a universal framework for users, planners, scientists and policy makers, (Medema et.al, 2008). Scholars have argued theoretically that IWRM promises to significantly contribute to solving some of the developing world's water management problems (Manthrithilake, 2008, and Bhaduri et.al, 2011). However, there have been criticisms of IWRM. Scholars have also pointed out that the IWRM concept in operational terms has not been sufficiently, comprehensively and critically evaluated (Biswas, 2008). There are further criticisms that there is a gap between IWRM as a theory and IWRM in practice (Shah, et.al. 2006; Rahaman et.al. 2005). Nonetheless, the IWRM is becoming the water resource management framework for river basins, but has yet to address the particular characteristics of lake basins, with characteristics such as integrating nature, long retention time, and complex response dynamics, resulting in challenging implications for governance (World Bank, 2005).

The idea of an integrated governance framework for one resource, such as water, is illogical, considering that water as a resource is naturally interconnected to other natural resources (Medema et.al, 2008). Currently, there is no consensus amongst scholars and practitioners on how best to understand and implement IWRM. Developers, governments and international organizations foresee problems with the IWRM through a lack of integration between water and other natural resource policies (Samuels,et.al. 2006)

The Adaptive Management Framework (AMF), also referred to as Adaptive Resource Management (ARM), or Adaptive Environmental Assessment and Management (AEAM) (Samuels, 2006) can be defined as a participatory and collaborative framework that has been utilised theoretically to understand resource governance rather than a framework used in practice. At the time of its implementation in the 1980's there were a number of barriers preventing its success. This was notable in the early part of the twenty-first century when AMF was inserted into global natural resource strategies and plans, particularly in the US, without a full appreciation of what resources are fully required to adequately use the framework. The barriers to its success included, institutional inertia, lack of organisational capacity, an absence of leadership and inadequate resources (Bormann et.al, 2006). Therefore, it is considered a theoretical framework that is still being unravelled, rather than a practical framework. In its earlier practical application, there was a misunderstanding with regard to what organisational resources were required to successfully implement the framework (Lee, 1999). Good Lake Basin governance requires adequate and sufficient human, material and financial resources. The lack of adequate resources, is a serious issue for the governance of Lake Chad Basin Commission, and any framework introduced to address its current issues could result in failure, if the lack of resources is not tackled. What is required, is investment in these resources that adequately meets globally recognised standards of good practice. One of the solutions that Lake Chad Basin Commission has implemented, is to work in partnership and outsource some of the key roles required to non-government organisations who can provide training and assistance on the ground.

In 1996, the World Bank endorsed a Lake Basin Management initiative to direct a focus to improve the overall governance of lake basin management (Ayers et.al, 1996). The initiative was based on a study of 28 lake basins in various geographic and socioeconomic locations around the world. The study of the 28 lakes confirmed that lake basins have a wide array of challenges, both from within and external to the lakes. As a result, the Integrated Lake Basin Management initiative was developed over a three-year period, 2009-2011, known as the African Lake Basin Management with Sanitation (AFSAN), which is currently being introduced in three lake basins in East Africa: Lake Nakuru (Kenya), Victoria (Kenya) and Lake Chivero (Zimbabwe).

According to the International Lake Environment Committee, there are seven key principles to good lake basin governance. First and foremost, a sustainable lake requires the financial, material and human resources that will enable a harmonious relationship between humans and nature. Other important principles include the sustainable planning and management of drainage system, long-term strategic approach to the prevention of lake degradation, the best information available to govern the lake, to improve strategic decision-making, resources and intelligence to address conflicts amongst competing users of the lake. Fairness, transparency and empowerment for all stakeholders, is also an important requirement. However, all of the initiatives to date have not adequately addressed the lack of financial, material and human resources that are required to enable good governance to take root.

The fourth effort to improve good standards of corporate governance and business practices within Lake Basin Management is the Environmental Impact Assessment (EIA). The EIA was initiated in the USA in 1969. The United Nations Environment Programme (UNEP) defines EIA as an environmental governance and management process that classifies the environmental, social and economic impacts of a proposed project, in advance of the decision-making stage. In other words, the EIA is a process that protects the environment by reducing adverse environmental impacts such as water pollution or contamination. The EIA improves the governance of natural resources. According to the UNEP the EIA identifies positive and negative environmental impacts, recommends programs of action to mitigate negative environmental impacts and improve positive effects. The EIA also includes a follow-up process of monitoring and environmental auditing. However, the process of monitoring and audit vary from country to country. The EIA is endorsed by a number of multilateral and bilateral agencies, including the Rio Declaration Principle 17, the United Nations Environmental Programme (UNEP), the World Bank, the OECD, and the United Nations Commission for Europe (UNECE). All four countries sharing the Lake Chad Basin have adopted the EIA. In 1992, Nigeria was the first of the four countries to adopt by establishing Environmental Impact Assessment Decree, No. 86. Three years later EIA Procedural Guidelines were introduced in Nigeria. The National Council on the Environment and the Federal Environmental Agency governed the whole process. In Cameroon, the EIA was implemented in 1996 as Environmental Management law no.96/12, with the manual guidelines for EIA introduced four years later. The Ministry of the Environment, Nature Protection and Sustainable Development governed this process. In Chad and Niger, the EIA was introduced in 1998 (Dalal-Clayton, et.al.1998).

17.2 How Lake Chad is attaining global standards of efficiency, commercial probity and effective economic and social development

To reinforce global standards of efficiency, commercial probity and effective economic and social development, the four Supreme Audit Institutions (SAIs) of the riparian countries of Lake Chad, the Supreme State Audit services of Cameroon, the Court of Auditors of Chad, the Court of Auditors of Niger, and the Office of the

Auditor General for Federal Republic of Nigeria, jointly coordinated an environmental audit of Lake Chad. The environmental audit was conducted to verify the following, firstly, whether the riparian countries were complying with global standards of efficiency and good practices, secondly, whether the process of control, monitoring and enforcement practices and systems were sufficient to achieve sustainable use of water resources in the Lake Chad Basin. And finally, whether the Lake Chad Basin Commission fulfils its role and responsibility, in relation to monitoring States' performances in the implementation of measures designed to control the Lake Chad water resources.

The findings of the joint environmental audit report found some good practices identified by each member state during the audit. In Cameroon, to reduce the impact of drought that affects the water resources of the Lake, the Ministry of the Environment, Nature Protection and Sustainable Development, carried out reforestation efforts in the region of the Lake Chad Basin, planting, just under three million tree seedlings, over 18,500 hectares of land between 2008 and 2013. This initiative was part of an operation known as 'Sahel vert'. In Niger, the Ministry of the Environment, Urban Health and Sustainable Development, conducted a number of land restoration, reforestation and land recovery actions. This action safeguarded and protected land in the lake basin. In Nigeria, the implementation of a coordinated shared charter for the management of water resources was established, in accordance with the Integrated Water Resource Management (IWRM). This charter promoted the importance of participative stakeholders. The charter was based on the following principles, firstly, recognising the rights of the local communities and indigenous people. Secondly, enabling a broad participation in the decision-making process, and thirdly, a clear definition and recognition of responsibilities and rights within the federal state. In Chad, good practice was identified in the implementation of activities for reforestation, as well as a ban on logging and poaching.

The lack of human, material and financial resources undermined other good practices. This is evident in the enactment of binding rules for water management and governance within the four riparian countries.

In each of the four riparian countries, binding rules to control the major uses of water resources have generally been enacted. For example, in Cameroon, on 14 April 1998, a law was enacted on water, marking a new era in the domain of water and sanitation. This law applied strict, national management and coordination of water resources, services and supply throughout the country, paying close attention to the protection of water and the liberalization of the supply sector. However, the auditors found evidence, that only two water inspection officers were responsible for ascertaining potential violations and performing relevant law enforcement efforts, for the Cameroon-section of Lake Chad. Two water inspection officers were deemed wholly inadequate and unsatisfactory to properly enforce the law, falling below standards of good practice, since the two officers did not have the necessary equipment for on-the-spot checks. Therefore, there was a lack of human and material resources to properly uphold the law on water use and regulation in Cameroon.

To address this issue, the audit team recommended a mobilization and redirection of adequate logistical human resources to enable annual investigations and checks on key water users. This action would prevent violations of water use and uphold the law on this issue.

On 1st April 2010, Niger enacted a Water Code. This code determined water resource management for the whole country, including the requirements for people, livestock and the hydro-agricultural developments. However, the auditors found that the human resources to enforce the Water Code were largely non-existent. There was no record of monitoring activity to enforce this law in Niger. The auditors recommended that Niger strengthen their technical, material and financial structures for monitoring, collecting data and enforcing sanctions provided by the binding rules and regulations. They also recommended the tightening and renovation of water resources to improve the monitoring and observations network and water use surveillance system.

In Nigeria, there were a number of binding rules enacted regarding the regulation and use of water.

These binding rules were to control all surface and groundwater and all watercourse in Nigeria. These rules were also used to promote good health and wellbeing of fauna and flora and promoting sustainable use of natural resources, including water. In 1987, the decree of the River Basin Development Authority was enacted followed by the River Basin Development Authority Act 1990 with an amended decree in 1997. Also in 2007, the Water Charter for Sustainable and Equitable Management of the Hadeja-Jama and Komadugu-Yobe Basin was enacted, as well as state water supply, environmental edicts and policies. However, the auditors found that the Ministries, Departments and Agencies within Nigeria, responsible for the control, monitoring and enforcement of such binding rules, did not allocate any resources for its execution. The auditors recommended that the Federal and State governments should establish a patrol team to enable regular inspection to ensure compliance with laws, regulations and standards. They also recommended proper enforcement coordination among all ministries, departments, agencies responsible for enforcement.

Chad enacted the Water Code, in August 1999. The Water Code was enacted to protect and safeguard water resources from pollution and contamination risks, to ensure safe drinking water, and to meet the needs of agriculture, animal husbandry, industry, transport and any other activities of interest. However, the auditors found that although the National Water Management Committee authorizes the responsibility of implementing water resource protection measures to ministries capable of water management activities, there was a lack of material and human resources that hindered these ministries from adequately carrying out their task. In some cases certain sites around the lake are not checked or monitored at all. Again, the auditors recommended that adequate financial, human and material resources are allocated each year to allow for the fulfilment of their mission regarding the control of water use.

The key failures of the Lake Chad Basin Commission governance system are partially explained by a lack of resources. Evidence from the environmental audit report confirms this. When the auditors verified that the medium to long-term strategic development plan of the Lake Chad Basin Commission, did not adequately link the actions and water needs of each country's local, national and regional needs. It was clear that identifying water needs locally, nationally and regionally requires human competence, skill and intelligence, as well as material resources in the form of tools and equipment, and the financial resources to enable such action.

The auditors also confirmed that there was inadequate enforcement of rules for the protection of water resources, that the systems, procedures, methods and tools for regulating the withdrawal of water resources of the Lake Chad basin in each state were not functional. The commitment from leadership in each state, to ensure the collection, processing, distribution and archiving of data, in relation to the evolution and use of water resources were not evident. This was accompanied by a poor unsustainable funding model, and the lack of effective actions for the sustainable management and protection of water resources of the Lake Chad basin. This evidence confirms the need for adequate investment to uphold and strengthen governance.

17.3 What institutions are in place that are advancing, teaching and disseminating standards?

There are three key institutions in place that are advancing, teaching and disseminating standards on environmental governance of natural resources within Africa. Firstly, we have the International

Organisation of Supreme Audit Institutions (INTOSAI), an umbrella organisation founded in 1953, with the general objective of good governance. It was established to support the transfer of knowledge, develop government auditing worldwide and improve professional capacity (INTOSAI, 2004). In 1976, the African Organisation of Supreme Audit Institutions was set up, as a regional branch of the INTOSAI, with the mandate to audit public finances within Africa, consisting of three main duties. Firstly, to implement organised studies in relation to the audit of public finances, followed by the responsibility to deliver and conduct training courses for all personnel, and finally to cooperate, pool resources and work with the International Organisation of Supreme Audit Institutions (INTOSAI). Today the INTOSAI has seven regional Supreme Audit Institutes, representing different parts of the world. The African Organisation of Supreme Audit Institution is one of those regional bodies. The importance of establishing and implementing environmental audit and its role in advancing, teaching and improving standards of environmental governance, did not emerge within African professional accounting circles until the early part of the 21st century. The term and development of environmental audit began in the early 1970's, during an era of environmental legislation, prosecution and civil law suits against corporations that were causing environmental damage to the environment, particularly corporations within the United States (INTOSAI, 2004). These corporations were using environmental audit as a methodological, systematic approach to addressing these issues. These issues were recognised, in the 1992 conference, in Rio de Janeiro, hosted by the United Nations, regarding the growing concern over environment matters. In response to this conference by the United Nations, the membership of the International Organisation of Supreme Audit Institutions (INTOSAI) agreed on the principles of setting up a working group on environmental auditing (WGEA) to address issues of environmental protection. This decision was made in Washington D.C. in 1992. Ten years later, in 2002, after conducting a number of environmental audits in South Africa, the African Supreme Audit

Institute (AFROSAI), approved the Working Group on Environmental Audit (WGEA) in Africa. The Working Group on Environmental Audit (WGEA) in Africa was mandated to develop as well as improve environmental audit standards, instruments and tools for environmental audit work, to prepare guidelines and information to improve practice and to further the understanding of specific environmental concerns (INTOSAI, 2004).

The Supreme Audit Institutes (SAI), all over the world, including Africa, have chosen to audit water issues as a central theme of their work, since it is vital to humanity, biodiversity and the environment. Their purpose in focusing on water issues is to improve the quality and standards of water management, promote higher standards of governance and improve public health and the environment. According to the (INTOSAI, 2004), a good starting point for an environmental audit is to review the national, regional, strategies, policies and programmes on water issues as well as the budget spent on water, which has been agreed and set by several government bodies in cooperation with private organisations. For example, the auditors would examine the generic environmental water policy, reviewing the purpose of the policy, the policy controls, the division of roles, responsibilities, authorities and tasks. Other important elements would include signed water policy agreements between government and executive bodies, as well as middle and lower managerial levels. Auditors would also verify the international agreements on compliance, in terms of water legislation, laws and legal instruments, making a comparison between the water legislation within Africa and that of international legislation, (INTOSAI, 2017).

Another important institution that is advancing, teaching and disseminating standards, is the United Nations Commission for Europe (UNECE). The Commission sanctioned the Convention on the Protection and Use of Trans-Boundary Watercourses and International Lakes (UNECE, 1999). This convention was approved in Helsinki Finland, on 17 March 1992, and implemented on 6 October 1996, by member States of the United Nations

Commission for Europe (UNECE) and regional economic integration organizations constituted by such States. On 28 November 2003, a decision was agreed, amending article 25 and 26 of the Convention, permitting all United Nations Member states to voluntarily comply with the terms of the Convention. This Convention exists to promote trans-boundary water cooperation as well as to protect the quantity, quality and sustainable use of trans-boundary water resources. It aims to strengthen governmental negotiations and agreements at bilateral and multilateral levels, increase mutual understanding and collaboration for the prevention, control and reduction of trans-boundary water pollution and sustainable use of trans-boundary waters, promoting good faith, friendliness and good cooperation between the riparian parties. There are a number of key benefits of trans-boundary water cooperation. The UNECE policy guidance notes communicated these benefits, which include accelerated economic growth, improved human well being, enhanced environmental sustainability and increased political stability (UNECE, 1992).

On a bilateral and/or multilateral basis, the parties and key actors in trans-boundary watercourses and international lakes, are encouraged to collaborate on education, research and development in international forums to improve knowledge and the exchange of information, capacity building, as well as advice for trans-boundary watercourses and international lake projects on the ground, with the aim of improving the condition of trans-boundary lakes, develop sound water-regulation techniques, examine and inspect the physical and financial damage to lakes, as well as improving methods of assessment on hazardous substances. The Convention also aims to develop and implement, comparable, legal, administrative, economic, financial and technical measures to reduce water pollutants, limit waste water, as well as the implementation of environmental impact assessments that can assist in reducing environmental risk (UNECE, 1992). There also exists a working group on integrated water resources management and the working group on monitoring and assessment. These are the two main subsidiary bodies established by the Trans-Boundary

Watercourses and International Lakes (UNECE, 1992). The main focus of the working group is to support the implementation of the Convention, strengthen the management of trans-boundary bodies and promote access to information and public participation in decision-making. The International Standards Organisation in Environmental Management 14001, is another institution that advances, teaches and disseminates standards in natural resource management and governance. The main purpose of this International Standard is to provide organisations with a charter to protect the environment and respond to varying environmental conditions.

17.4 What support have the riparian parties received to promote good governance?

In general, many Commonwealth countries have received support from the World Bank, African Development Bank, the United Nations and many other international institutions to promote good governance. The riparian parties surrounding Lake Chad have to some degree received support financially for projects and programmes in connection with water resources management, as well as support in the form of the development of policies, projects and programmes affecting agriculture, the environment and many other issues in connection with Lake Chad Basin. However, it is questionable, particularly in the light of the recent AFROSAI joint environmental audit report, whether they have been strengthened in their capacities to maintain and extend good practice in good governance. The German Agency for International Cooperation for Development (GIZ), throughout the period 2011 to 2016 with plans to extend to 2018, have been investing in the institutional reform of the Lake Chad Basin Commission. This reform programme includes planning, human resource processes, monitoring and evaluation and information systems. The management information system has been modified and adjusted to meet the current challenges that confront the Lake Chad Basin Management. These current challenges will also take note of the recommendations made by the recent environmental audit of Lake Chad. The success of the GIZ programme is yet to be realised.

The Global Environmental Facility (GEF), established in the 1992 Rio Earth Summit, works with the objective to tackle the world's highest prioritized environmental issues. Over a 4-year period, starting from 2003, GEF started a project titled, 'Reversal of Land and Water Degradation Trends in the Lake Chad Basin Ecosystem.' The purpose of the project was to overcome barriers to the management of the basin through enhanced collaboration and capacity building. However, the GEF assessment team found that stakeholder participation in the pilot programme was low and not representative. There was no clearly identified method or approach to guarantee indigenous participation. There were recommendations to have traditional leaders to represent and mobilise the indigenous population; however, not all community groups had a traditional leader to represent them. They also found that there was an open access for all, to natural resources in the Lake Chad basin, with no environmental management system in place to restrict or monitor the usage of water resources. There was no wetlands management, ground water management or watershed management. However, there were some good examples of community-based fisheries management by three villagers, in the region of Chad. The outcome of the project was to expand the pilot study into a full programme of study. The World Bank was one of the major financial investors to support the Lake Chad development and Climate Resilience Action Plan (LCDAP). The project was approved in 2015, part of the World Bank's Africa Climate Business Plan, presented at COP21 climate change conference. The total investment is projected at 1 billion US dollars, with seven priority themes, including, better management of the water resources of the basin, being one of the priority themes as well as securing access to natural resources and managing conflict. The African Development Bank has also invested in a programme that calls for international cooperation to safeguard Lake Chad. The programme is titled, 'To rehabilitate and strengthen resilience of Lake Chad Basin Socio-

Ecological Systems'. The programme has three key objectives, (i) to strengthen the socio-economic system, (ii) to improve adaptation to climate change, (iii) to improve peace and sound governance of shared resources. The total cost of the programme is $71.23 million. The project will also collect necessary satellite agricultural, climatological, limnometric and socio-economic data. The outcome of these projects is yet to be realised but the return on investment looks promising.

17.5 What impact have global developments in corporate governance had on the governance of Lake Chad Basin?

There are a number of key global developments in corporate governance that have influenced and shaped the development of natural resource governance in the Commonwealth as a whole, and in the Lake Chad Basin. In particular, these developments in corporate governance include the need for sustainable economic growth, where the economic growth of nations should not be at the expense of humanity, biodiversity and the eco-system. For example, the current joint economic growth of the United States of America, China and India is unsustainable. The global economy as it stands needs to limit, restrict and reduce the amount of manufactured production, consumption in the form of buying and selling of products that harm the environment, particularly at their end of life cycle, resource depletion, and waste generation to levels that planet earth can reasonably absorb without threatening current and future generations of humanity, biodiversity and the eco-system. These developments in corporate governance, with particular reference to natural resource governance, has swayed corporations in developed countries (global north) to conduct their affairs in developing countries (global south) with reference to higher board accountability in sustainability disclosure, risk management, stakeholder engagement, culture and ethics. These changes have resulted in the transformation in the organisational approach and behaviour notably to environmental control and risk. These developments have led to the adoption of risk based approaches to the management of natural resources in terms of adopting a risk culture and robust risk based process approach, particularly in supply chain management, in terms of ensuring the origins of natural resources utilised by an organisation in the global north is sourced in line with good trading practices and in compliance with trade certification schemes, with the objective of promoting good governance in natural resource management. Natural resources such as oil and gas, minerals, timber and water have all been subject to global certification standards and procedures by the global north, that have influenced good governance practices in Commonwealth countries. For example, the International Standards Organisation (ISO) have issued a number of international standards in relation to water, these include water quality standards (ISO/TC147). This standard provides guidance on water sampling methods, reporting and monitoring of water, to determine impurities in water quality, such as bacteria. The ISO in water footprint (ISO14046) provides guidance on the measurement and reporting of water footprint, setting an international standard and benchmark for water usage, ISO (2017). ISO, in Hydrometry, includes 70 standards for measuring, planning and water conservation for ground water. There are three International Standards on drinking water supply and waste water (ISO 24510, ISO 24511 and ISO 24512). These standards provide guidelines for water authorities and operators to assess and improve the delivery of drinking water and waste water systems. There are also a number of international standards that provide guidance for crisis water management, (ISO 24518, ISO 24520, ISO24527 and ISO 24522), ISO (2017).

The global north has higher standards of natural resource governance, through their certification schemes, standards and supply chain processes, that are influencing the commonwealth approach to such practices, which in turn influences commonwealth approach to natural resource governance.

17.6 Concluding remarks

Currently, the governance of Lake Chad Basin Commission lacks human, material and financial resources to adequately meet international standards of good practice in corporate governance. This statement is supported by the recent environmental audit report conducted by the African Supreme Audit Institute.

REFERENCES

Ayers, W.S., A. Busia, A. Dinar, R, R. Hirhi, S.E. Lintner, A.F. McCalla and R. Robelus (1996). *Integrated lake and reservoir management.* World Bank Technical Paper No. 358.

Biswas, A. K., Braga, B. P. F., Tortajada, C. and Palermo, M. (2008). 'Integrated water resources management in Latin America'. *International Journal of Water Resources Development,* 24(1) special issue.P.2 (online) URL: http://dx.doi.org/10.1080.07900620801902918.

Bhaduri, A..Hornidge, A.K., Manschadi, A.M., Oberkircher, L., Schorcht, G., and Tischbein, B. (2011). 'Reconceptualizing water management in Khorezm, Uzbekistan'. *Natural Resource Management Forum.* P.5

Bormann, B.T., Clark; R.N. and Stankey, G.H. (2006). 'Learning to manage a complex ecosystem' *Adaptive management and the Northwest Forest Plan.* P3

Chad (1998) Law no.14 / PR / 98 defining the general principles of the protection of the environment.

Dalal-Clayton, B., Donelly, A. and Hughes, R. (1998). 'A Directory of Impact Assessment Guidelines'. Published by the International Institute for Environment and Development, 2nd ed. P.55,61

Dinar, A. and Saleth, R.M. (2005). Water institutional reforms: theory and practice. *Water Policy* , P.8.

Global Water Partnership (GWP) (2000). *Towards Water Security: A Framework for Action.* Published by the Global Water Partnership. P.14

Hubschen.K. (2011). '*Integrated Water Resources Management as a Governance Challenge for Countries of the Middle East with Special Focus on Yemen, Jordan and Syria*'. Published by Logos Verlag Berlin, P,3.6.

INTOSAI (2004). 'International Organization of Supreme Audits: 50 years (1953-2003). A Special Publication of the *International Organization of Supreme Audit Institutions'*. P.18-96 (online)

http://www.intosai.org/uploads/412efestschrift.pdf

ISO (2017). 'ISO and water. Great things happen when the world agrees'. Published by *International Standards Organisation.* P.9-13 (online)

https://www.iso.org/files/live/sites/isoorg/files/archive/pdf/en/iso_and_water.pdf

Joint Environmental Audit (2014). 'Joint Environment Audit on the Drying up of Lake Chad. Joint Audit Report'. Published by the *Deutsche Gesellschaft für Internationale Zusammenarbeit (GIZ) GmbH Registered offices Bonn and Eschborn.* P.19-80.

Lee, K.N. (1999). 'Appraising Adaptive Management' *Conservative Ecology*, 3(2):3, [online]

http://www.consecol.org/vol3/iss2/art3.

Leeuwen, S.V. (2004). 'Auditing Water Issues, experience of Supreme Audit Institutions'. Published by *International Organisation of Supreme Audits Working Group on Environmental Auditing* (online)

http://revista.tcu.gov.br/ojs/index.php/RTCU/article/viewFile/656/1363.

MacKenzie, Susan H. *1996*. 'Integrated Resource Planning and Management: The *Ecosystem* Approach in the Great Lakes Basin'. Washington, D.C: Island Press. P.3.

Manthrithilake, H. (2008). 'Land and water management issues in Central Asia'. Presentation at the international conference on the problems of the Aral, their influence on the genefund of the population, vegetation and animal world and measures of international cooperation to mitigate their consequences, Tashkent, 12 March.

Medema, W., McIntosh, B. S. and Jeffrey, P. J. (2008). 'From premise to practice: a critical assessment of integrated water resources management and adaptive management approaches in the water sector. Ecology and Society 13(2): 29. [online] URL: http://www.ecologyandsociety.org/vol13/ iss2/art29/.

NNPC (2017). Nigerian National Petroleum Corporation, 'Investment potential' (online)

http://www.nnpcgroup.com/NNPCBusiness/BusinessInformation/OilGasinNigeria/InvestmentPotential.aspx

Odada, E., Oyebande, L. and Oguntola, J. (2006). 'Lake Chad: Experience and lessons learned brief. (online) http://www.ilec.or.jp/eg/lbmi/pdf/06_Lake_Chad_27February2006.pdf (accessed on 11/12/2013).

Rafik, H. (2009). 'Lessons for Managing Lake Basins: Measuring Good Governance'. Water P-Notes; No.25. World Bank, Washington, DC. World Bank, P.2 (online) https://openknowledge.worldbank.org/handle/10986/11735.

Rahaman. M.M. and Varis, O. (2005). 'Integrated water resources management: evolution, prospects and future challenges. Sustainability: Science, Practice & Policy, P.17.

Ramsar (2006). 'Two new Ramsar sites in Chad. Chad's newest Wetlands of International Importance'. (online) http://www.ramsar.org/news/two-new-ramsar-sites-in-chad

Samuels, P., Woods, B., Hutchings C. (2006). 'Sustainable water management in land-use-planning'. *CIRIA*, London. P.203

Shah, T. and van Koppen, B. (2006). 'Is India ripe for integrated water resources management? Fitting water policy to national development context'. Economic and Political Weekly XLI (31): 3413-3421.

UNECE (1992). 'Convention on the Protection and Use of Trans-boundary Watercourses and International Lakes. UNECE P.14. (online)

https://www.unece.org/fileadmin/DAM/env/water/pdf/watercon.pdf

World Bank (2005). Lessons for Managing Lake Basins. Measuring Good Governance. Published by the World Bank.

Endnotes

---

[197] There are other countries that have gained indirect access to the Lake but are not included in this chapter, only those countries with direct access and contact to the waterbodies of Lake Chad are discussed.

Chapter 18

Corporate Governance in Commonwealth Countries: Concluding Remarks and Recommendations

18.0 Concluding Remarks

The need for good governance is now internationally recognised. This is because of the high correlation between corporate governance and investor decisions[1]. If the management of the economic and social resources for development in a country is efficient and there is transparency and accountability in the government, there is likely to be high levels of private sector investment inflows into the country. Good corporate governance improves the financial performance of companies and the capital markets. Within the CW, the approach is to transform corporate governance into an essential part of the policy environment and institutional infrastructure of every country. As countries face the global market, national economic policy makers will need to understand and to be aware of the effect of poor corporate governance on companies. In many developing countries where a significant proportion of previously state owned companies have been privatised, and where increasing numbers of companies are being established, the private sector would seem to hold the key to the economic growth and development of these countries. Governments in the developing countries of the CW must create an enabling environment for the private sector to spearhead economic development while at the same time encouraging the participation of all stakeholders in development. There is the need to extend and apply corporate governance principles to a wide range of organisations, and not just the publicly listed joint stock companies.

Such organisations include the banking sector, state enterprises, co-operatives, and the growing and increasingly important NGO sector and in public services such as health and education boards. Efficiency, transparency and accountability are essential to achieve sustainable development. This means that the impact of improved corporate governance can be felt in all sectors of the economy.

Founded in 1998, the CACG played an appreciable role in promoting good standards of corporate governance within the CW; this is despite the differences of culture, wealth and size of member countries. It mounted corporate governance initiatives to build capacity in over 30 countries, and produced the Commonwealth Principles for Corporate Governance. It also produced the Best Practice Guidelines for Boards of Directors, the Kenya Code of Corporate Governance and Guidelines for Directors, and the Manual for Directors. This was possible because of a platform of essential commonalties – 'the Commonwealth Factor'. Whilst the member countries of the CW share a number of commonalties, it is important to recognise that any principles of corporate governance recommended must be such that implementation can be adapted to the national and company specific circumstances. Beyond drafting and adopting these principles, is the need for implementation and action. Circumstances and priorities sometimes differ between countries and regions, and 'one size, may not fit all' when it comes to implementation. Even if there is an increased global understanding concerning basic criteria for good corporate governance, the approach to regional co-ordination has to be adapted to this. Capacity building and regional co-ordination has to take account of specific circumstances. It should also make full use of existing and ongoing regional initiatives, including already developed corporate governance codes and guidelines. There are already several examples of such initiatives in Africa and the CW.

Hence Demirag et.al. (2000) suggest that any evaluation of the effectiveness of the corporate governance framework in a country requires not only a mapping of the range of governance mechanisms available but also the identification of the potential interaction among them. They state that the evaluation of any single mechanism, which ignores the interaction, is "incomplete and may lead to untenable and erroneous policy formulations to correct alleged governance failure" (2000:348). Accordingly, it would appear that whilst the effort and the achievements of the CACG in promoting good standards of corporate governance within the Commonwealth were commendable, they suffered from some drawbacks because of the failure to address the interaction between the range of governance mechanisms in CW member countries. As a former Secretary-General of the Commonwealth acknowledged (CBC, 2000/2001), "while globalisation has brought its benefits, there is a sharp edge for those inadequately orientated to ride its wave". According to him, there is a growing divide between developed and developing countries within the CW. Many of the emerging economies within the CW lack the institutions and resources necessary for good corporate governance. In these countries, the business environment lacks many of the elements required for a competitive market and the culture of enforcement and compliance found in the developed world. This is evident from the review of corporate governance developments in many of the countries presented in this book.

Furthermore, the reform of corporate governance systems in the environment of many of the emerging and transition economies can be politically difficult. According to Wakaisuka-Isingoma and Nkundabanyanga

> "Uganda has made significant strides in corporate governance. However, there are still a number of shortcomings that need to be addressed. For example, some laws are not fully aligned with international standards and political interference in the administration of justice, compounded by pervasive corruption, undermines the Ugandan investment climate. Corporate financial transparency is largely lacking with the exception of banks, insurance, and listed companies".

This assertion is true of many of the African member countries of the CW, which are fraught with political, religious and ethnic divides. There is also the problem of institutionalised corruption. Corruption is an enemy to good corporate governance and is an issue which member countries of the CW, especially the developing economies must address if the benefits of good governance are to have any effect at all. Whilst corruption is not unique to the developing countries, nevertheless, it has crippling effects on these countries because these economies are small and vulnerable to global vagaries.

The Commonwealth supports its member countries to progress towards sustainable management and regulation of their natural resources. It helps to strengthen policies, legal frameworks and national institutions providing advice and guidance to boost investment and enhance good governance in natural resource management. However, the Joint Environmental Audit Report (2014) of the Supreme State Audit Services of Cameroon, Chad, Niger and Nigeria, over a three-year period (2011-2014) found serious corporate governance failures in the management of Lake Chad by the Lake Chad Basin Commission. Such governance failures have arisen from, amongst others, the lack of financial, human and material resources.

Evidence gathered from the corporate governance developments in the CW member states presented in this book, reveals some positive outcomes:

- Many of the CW member states are making concerted efforts to address their corporate governance challenges
- Majority of them have legislation in place, including codes of corporate governance, which have been revised to adhere to international standards of best practices

- Many of the countries have revised their outdated company legislation in response to the changing global economic climate
- Majority of the countries have received support from international organisations and agencies, including the United Nations Conference on Trade and Development (UNCTAD), the Organisation for Economic Cooperation and Development (OECD), the Committee of Sponsoring Orgnisations (COSO), the International Finance Corporation (IFC), the World Bank, and many others.
- Some networks such as the African Union (AU), African Corporate Governance Network (ACGN), African Peer Review Mechanism (APRM), and Asian Corporate Governance Association (ACGA) are helping to promote good corporate governance practices in countries within their regions.
- The small states in the Pacific region are ably supported by the governments of Australia and New Zealand, in addition to support received from other international agencies
- Countries like Kenya, Rwanda and Mauritius are achieving positive results in their efforts to tackle corruption

And some shortcomings, including:

- The lack of political will to effectively address corporate governance in some developing countries
- Challenges with economic environment particularly the pervasive levels of corruption and perceived weak and ineffective judiciary, which undermines the rule of law, and thus poses a significant barrier to doing business

- Poor/inadequate business environment as revealed by the World Bank *Ease of Doing Business* data. Many of the CW member states do not have the enabling environment for attracting foreign capital because of the administrative bottlenecks that restrict potential investors.
- The environment for enforcing compliance with corporate governance and other company legislation is often weak. This creates conditions where the concentration of share ownership has become the norm among corporations as a way to maintain shareholder control in the absence of an active market and enforcement mechanisms. The lack of the dynamic market and implementation mechanisms creates an adverse legal environment for the protection of minority shareholder interests and in turn, further, discourages the formation of the private-sector investment.
- Political influence in the constitution and operations of public sector boards, which may affect their effectiveness.
- Most of the companies are either family owned or owner-managed businesses that perceive corporate governance requirements as costly and unnecessary, since agency problems do not exist. These companies are characterized by complex control structures, secretive shareholder agreements and severe potential conflicts of interests.
- Most boards consist of nominee directors, which are often selected to board memberships on grounds other than their professional competence. Such boards further consist of very few independent directors. As a result, board independence is distorted. This is the situation that is prevalent amongst most state-owned enterprises, government companies, parastatal organizations and the executive agencies.
- The lack of adequate human, material and financial resources for the effective governance of Lake Chad Basin.

18.1 Recommendations

Corporate governance is essential to contribute to the development of the private sector in the CW member states, which is currently viewed as the "engine of economic growth." Development of the public and private sectors depends, among other things, on the steady supply of long-term capital for which the effectiveness of corporate governance is a critical factor. Developing countries face the challenge of attracting the foreign capital flow needed to stimulate their economies. Improving corporate governance contributes to improving a country's competitiveness in attracting foreign capital flow. Therefore, these countries must be ready and willing to nib corruption at the bud and take the steps necessary to address their corporate governance shortcomings. Countries such as Mauritius and Rwanda provide great examples of how political will to enforce good governance can bring about positive change, which will significantly influence the economic landscape of the country. Appropriately, if CW member states are to develop their economies, they need good governance and enhanced integrity, which are pillars for better economic and investment climate. In the Corruption Perception Index scores examined earlier, Transparency International reports that many countries lack the political will to counter corruption. Although there is the notion that a number of countries are addressing the issue of corruption, the impact of any real change in these countries is yet to be felt. Therefore, the following recommendations are put forward:

- Leadership from the top and by example, and a zero-tolerance approach to corruption, as in Botswana, is needed, in many of the developing economies in the CW. Senior Management mainly politicians and professionals have to institute and promote good governance and create an environment for integrity as a foundation for creating a favourable investment climate. When the government operates in an environment of integrity, it creates trust and the willingness of the business community to cooperate. Such cooperation is of utmost importance in maintaining integrity (Fulgence 2014). Without strong, clear leadership and support from top government officials, its various departments and law enforcement agents, it is tough for any real advance in the fight against the evils of corruption to succeed (Sitta 2005; Fulgence 2013). Therefore corporate governance has to be at the top of the management agenda as a pro-business policy and should be embraced by all stakeholders in the society (Sitta 2005; Fulgence 2013; 2014).

- There is the need for programs to create public awareness and promote effective corporate governance. The information regarding corporate governance adherence and the consequences of non-compliance should be spelt out clearly in the client charters and disseminated freely (Sitta 2005; Melyoki 2005). Success in creating an environment of integrity for investment promotion and economic growth depends on the commitment and political will to implement necessary reforms as recommended (Fulgence 2013).
- Corporate governance reforms should take into consideration the needs of other stakeholders, other than the shareholders, as it is the case in Mauritius and South Africa.
- CW member states should consider adopting an integrated form of reporting, which will make directors more accountable for their disclosures in annual reports.
- Improve the strategic direction of state enterprises by providing appropriate training for the top executives in these countries to make them compete effectively in the global market;
- Grant international recognition to countries who have in place good corporate governance standards as priority locations for investment;

- Ensure high standards of corporate citizenship and business ethics among national companies and foreign direct investors in all participating countries;
- In many of the developing countries in the CW, family-owned small and medium sized enterprises (SMEs), constitute a greater percentage of the private sector. There is the need for adequate support for this sector, as they mostly depend on banks for their short and long-term financial needs, and may be subject to scrutiny by financiers as a condition for granting them loans.
- There is the need for the continuous support of smaller states, through the provision of training and financial resources, to enable them adopt good corporate governance practices.
- There is the need for continued investment to improve and strengthen the governance, environmental management and data monitoring of Lake Chad Basin, for the sustenance of all that rely on the resources the lake provides.
- There is the need to encourage and support research initiatives in corporate governance in CW member states.
- The Commonwealth Association for Corporate Governance (CACG) played an important role in promoting corporate governance in CW member states after it was established in 1998. However, it is no longer listed amongst the organisations and institutions accredited by the CW. Given the important role that the CW plays in supporting member states in different areas such as, health, education, sports, youth and women development and so on (see http://thecommonwealth.org/organisation-directory?page=13), there is the need to revamp the CACG. Corporate governance is too important a subject not to have its pride of place amongst institutions accredited by the CW. The CACG can support ongoing corporate governance initiatives in different CW countries, and provide a forum

for meeting and addressing corporate governance challenges in CW countries.

REFERENCES

Armstrong, P. (2000) "Corporate Governance in Commonwealth Countries" in *The Commonwealth Trade & Investment Almanac 2000-2001*, 58-63.

Commonwealth Business Council (1999), *The Commonwealth Trade & Investment Almanac 1999*. London: Hanson Cooke Ltd.

Commonwealth Business Council (2000), *The Commonwealth Trade & Investment Almanac 2000-*

2001. London: Kensington Publications Ltd.

Demirag, I., Sudarsanam, S. & Wright, M. (2000) "Corporate Governance: Overview and Research Agenda", *British Accounting Review* (32): 341-354.

Fulgence SE. (2013). Determinants of Corporate Governance Internet Reporting, LAP LAMBERT- Academic Publishing, Deutschland Germany.

Fulgence SE. (2014). 'Corporate Governance in Tanzania": In Idowu S.O and Caliyurt (Eds); *Corporate Governance; an International Perspective*. Berlin Heidelberg Springer; 157 – 185.

Melyoki LL (2005). Determinants of Effective Corporate Governance in Tanzania, Ph.D. Thesis University of Twente, Enschede. Netherlands ISBN 90-365-2160-2

Sitta SJ (2005). Integrity environment and investment promotion "The case of Tanzania" (unpublished) *Paper Presented to the Conference "Alliance for Integrity – Government & Business Roles in Enhancing African Standards of Living"*; Co-organised by the OECD, NEPAD, UN Global Compact and

Transparency International, in Addis Ababa on 7-8 March 2005

INDEX

A

B

C

F

G

H

J

T

U

V

W

Z

Made in the USA
Monee, IL
05 November 2026

67807083R00262